高职高专系列教材·计算机类

计算机专业英语

Computer English

主　编　王　莹　王　月
副主编　戴　赟　李世勇

西安电子科技大学出版社

内容简介

本书是按照高职高专人才培养目标和教学改革的实际要求，严格遵守“实用为主、够用为度、应用为目的”的高职高专人才培养的基本原则，根据《高职高专教育英语课程教学基本要求(试行)》编写的。全书分为8个单元，主要包括计算机硬件、软件、网络、多媒体、电子商务、计算机安全、大数据、云计算等内容。本书内容全面，集先进性、实用性、适用性和新颖性于一体。

本书可作为高等职业院校英语专业学生的教材，也可供电大等各类成人院校及广大专业人员学习专业英语使用，旨在提高计算机专业和相关专业学生以及IT行业从业人员的英语听、说、读、写、译的能力和英语交际能力。

图书在版编目(CIP)数据

计算机专业英语/王莹，王月主编. —西安：西安电子科技大学出版社，2018.4(2021.3重印)
ISBN 978-7-5606-4888-0

Ⅰ. ①计… Ⅱ. ①王… ②王… Ⅲ. ①电子计算机—英语—教材
Ⅳ. ①TP3

中国版本图书馆CIP数据核字(2018)第053980号

策　　划　高　樱
责任编辑　卢　杨　阎　彬
出版发行　西安电子科技大学出版社(西安市太白南路2号)
电　　话　(029)88242885　88201467　　邮　　编　710071
网　　址　www.xduph.com　　电子邮箱　xdupfxb001@163.com
经　　销　新华书店
印刷单位　陕西天意印务有限责任公司
版　　次　2018年4月第1版　2021年3月第6次印刷
开　　本　787毫米×1092毫米　1/16　印张　13.5
字　　数　319千字
印　　数　16 001～20 000册
定　　价　33.00元
ISBN 978-7-5606-4888-0/TP
XDUP　5190001-6

前　言

目前，我国高职高专教育已经进入“以加强内涵建设、全面提高人才培养质量为主”的全新阶段。为实现这一人才培养目标，高职院校英语教学改革正在如火如荼地进行着。作为改革的核心，课程内容的更新势在必行；作为内容的载体，教材建设至关重要。

在计算机领域，技术发展日新月异，大量的新思维、新概念不断涌现，诸如云计算、云存储应用等已相当普遍，大数据运营也已超出了传统的数据库管理系统的处理能力。“计算机专业英语”课程是计算机类专业学生学习专业知识的工具和桥梁，是满足 IT 岗位对人才的需求以及学生未来职业发展要求的有力保障。

依据国家对高职英语教学改革的要求，作为高职行业英语系列教材之一，本教材旨在帮助学生打好语言基础，同时培养学生的英语听、说、读、写、译的能力，尤其是提高学生未来工作中的英语交际能力。本教材由教改一线教师执笔，编写过程中得到了计算机行业专家的指导，保证了教材的质量和针对性，特点鲜明。

1. 先进性

以“服务发展”的职业教育理念为指导，以培养学生未来工作所需要的英语应用能力为目标，从而达到“推动高职英语教学与经济社会同步发展”的要求。

2. 实用性

内容选择上，以 IT 领域日常广泛应用的常识为前提；知识结构设计上，以英语语言内部规律为特点安排章节及练习，全面提高专业英语水平。

3. 适用性

“实用为主，够用为度”，充分考虑了目前高职学生的水平和可接受能力。

4. 新颖性

从听、说、读、写、译各方面全面设计教学内容，打破了“行业英语”以翻译为主的编写模式。

本教材分为八个单元，包括计算机硬件、软件、网络、多媒体、电子商务、计算机安全、大数据、云计算等方面的内容。每个单元均包括口语训练、听力训练、阅读及写作四大模块。

Speaking　口语训练图文并茂，深入浅出，引导学生了解行业全景、培养学生的职场交际能力。

Listening 听力部分涉及社会交往的各个方面，旨在帮助学生了解英语语言文化、学习社交技巧，为他们今后参与职场竞争打下基础。

Reading 阅读文章均选自计算机专业领域。精读课文后均附有计算机专业词汇以及练习题，使学生能够更好地理解原文，掌握所学要点。泛读课文涉及学生感兴趣的热点话题，能够抓住学生眼球，达到了提高学生阅读专业资料能力的目标。

Writing 写作部分包括应用文的介绍、写作模板及相关练习，目的在于培养学生书写商务信函的能力。

本书第1、2、7、8单元及相应附录由辽宁水利职业学院王莹编写，第3、4、5、6单元及相应附录由辽宁水利职业学院王月编写，王莹对本书进行了统稿。江苏省常熟职业教育中心校的戴赟老师也参与了相关编写工作。

由于时间仓促，水平有限，编写过程中难免有纰漏和不足，恳请各位专家和读者批评指正，并将相关意见和建议反馈给我们，以便再版时修订。

编者

2017年11月

Contents

Unit One Computer Hardware

Unit 1

Section One Speaking

Task I

1. Look at the following graphic and speak out each part of computer hardware system in Chinese.

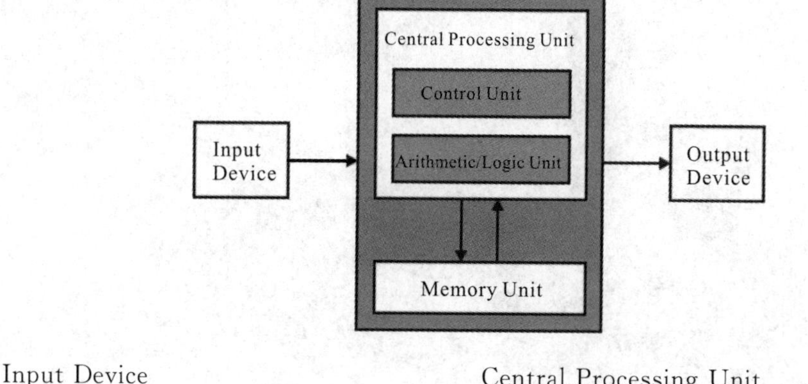

Input Device __________ Central Processing Unit __________

Control Unit __________ Arithmetic/Logic Unit __________

Memory Unit __________ Output device __________

2. List the Chinese names of the marks in the following hardware components of a personal computer.

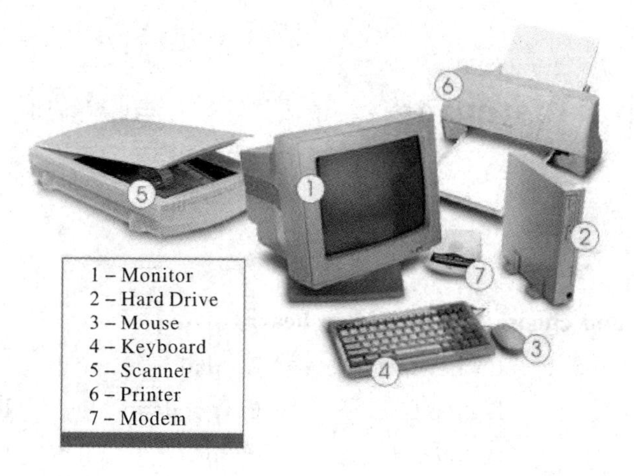

3. Suppose you are a computer engineer. Give a presentation to each picture in English.

Picture 1

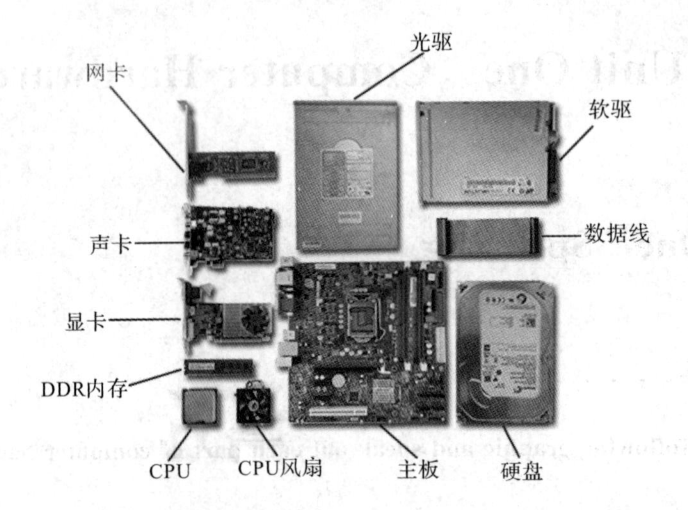

Picture 2

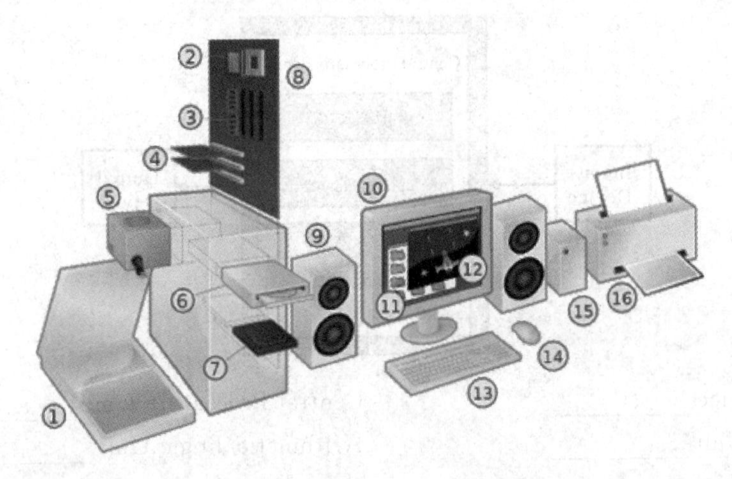

Hints: Basic hardware components of a modern personal computer including a monitor, a motherboard, a CPU, a RAM, two expansion cards, a power supply, an optical disc drive, a hard disk drive, a keyboard and a mouse.

Section Two Listening

Task Ⅱ

1. Listen carefully and choose the words you hear.

(1) A. bat	B. bet	C. dad	D. debt
(2) A. mood	B. food	C. good	D. look
(3) A. peace	B. please	C. piece	D. pleasure

(4) A. shape B. sheep C. ship D. sleep
(5) A. cheese B. cheap C. chair D. chemical
(6) A. form B. fur C. firm D. further
(7) A. pig B. bit C. beat D. dig
(8) A. goods B. glass C. guess D. class
(9) A. board B. Bob C. boot D. boat
(10) A. deep B. kid C. keep D. kick

2. Listen to the short passage and choose the proper words to fill in the blanks.

American and __1__ (British, Chinese, Japanese) people use different greetings. In the USA, the commonest __2__ (saying, greeting, speaking) is "Hi ". In Britain it is "Hello!" or "How are you? ". "Hi! " is spreading into British, too. When they are introduced to someone, the Americans say, "Glad to __3__ (meet, know, see) you ". The British say, "How do you do? " or " __4__ (Glad, Nice, Pleased) to meet you. " When Americans say "Goodbye ", they nearly always add, "Have a good day. " or "Have a good trip. " etc. to friends and strangers alike. The British are already beginning to use "Have a good __5__ (day, trip, holiday). "

3. Listen to the dialogues and choose the best responses to what you hear.

(1) A. What a pleasure.
B. I'm very pleased.
C. It's my pleasure.
D. Pleased to meet you. ______

(2) A. Thank you. I'm Li Guang.
B. Hope to see you again.
C. How are you doing?
D. Glad to meet you. ______

(3) A. What brings you here?
B. How do you do, Miss Evans? Very glad to see you.
C. Very good. How are you?
D. Hi, Lucy. What a nice surprise! ______

(4) A. How are you, Tom?
B. Fine, thank you.
C. How do you do, Tom?
D. Not so bad. ______

(5) A. I'm glad to meet you.
B. How old are you, Sun Hao?
C. Is that so, Tom?
D. Are you really Sun Hao? ______

(6) A. Say hello to you.
B. How are you doing?

C. How do you do?

D. It's a pleasure to meet you. ________

(7) A. Nice to meet you, Helen.

B. How are you getting along?

C. Fine, thanks.

D. Thanks a lot. ________

(8) A. That's very nice.

B. Nice to meet you, too.

C. It's very kind of you.

D. So do I. ________

4. Listen to the dialogues and choose the best answer to each question you hear.

(1) A. Mary. B. Jane. C. Sue. D. Kitty.

(2) A. Spring. B. Summer.

C. Autumn. D. Winter.

(3) A. Play football. B. Play basketball.

C. Go to the cinema. D. Go cycling.

(4) A. By taxi. B. By bus.

C. On foot. D. By underground.

(5) A. At a quarter to three. B. At three o'clock.

C. At a quarter past three. D. At half past two.

5. Listen to the dialogue and answer the following questions by filling in the blanks.

(1) Where did the dialogue take place?

It took place ________.

(2) What is Liu Wei?

He is ________, who is here to meet Jenny.

(3) What did Jenny think of the trip?

She thought it was not ________ on the whole.

(4) Why did Jenny feel tired?

It's ________ to China.

(5) How long did it take Jenny to fly to China?

For about ________ hours.

6. Listen to the passage and fill in the blanks with words you hear. Some new words given below will be of some help to you.

seem [siːm] v. 好像

mathematics [ˌmæθɪˈmætɪks] n. 数学

weekday [ˈwiːkdeɪ] n. 工作日

climb [klaɪm] v. 爬

mountain [ˈmauntɪn] n. 山

picnic [ˈpɪknɪk] n. 野餐

graduate [ˈgrædjueɪt] v. 毕业
graduation [ˌgrædjuˈeɪʃən] n. 毕业
celebrate [ˈselɪbreɪt] v. 庆祝
friendship [ˈfrendʃɪp] n. 友谊
cherish [ˈtʃerɪʃ] v. 珍惜
dream [driːm] v./n. 梦想；梦
arrive at 到达
have a good time 玩得开心

My Experiences in College

Like many other freshmen, I showed great __1__ in college life when I arrived at the college. I saw new faces, made new __2__, went to new places, and had new classes. Everything seemed new and interesting to me.

One week later, we began to have __3__. On weekdays, I had many classes, such as mathematics, business and __4__. On weekends, I would climb the mountains or go to the __5__ with my classmates by bicycle. Sometimes we had a __6__ and I usually had a good time there. When we were about to graduate, we held a lot of __7__ on campus to celebrate our graduation and show our __8__ to each other.

Now five years have passed, but college life is still __9__ in my mind and I still cherish those good old days at college. I miss my classmates very much, and I dream that all of our classmates can __10__ again at college some day.

Section Three Reading

Passage A

Computer Hardware (Ⅰ)

Computer hardware is the collection of physical components that constitute a computer system. Computer hardware is the physical parts or components of a computer, such as monitor, keyboard, computer data storage, graphic card, sound card, motherboard, and so on, all of which are tangible objects. By contrast, software is instructions that can be stored and run by hardware.

Hardware is directed by the software to execute any command or instruction. A combination of hardware and software forms a usable computing system.

Personal computer

A personal computer, also known as the PC, is one of the most common types of computer due to its versatility and relatively low price. Laptops are generally very similar, although they may use lower－power or reduced size components, thus lower performance.

Computer case

The computer case is a plastic or metal enclosure that houses most of the components. Those found on desktop computers are usually small enough to fit under a desk; however, in recent years more compact designs have become more commonplace, such as the all-in-one style designs from Apple, namely the iMac. A case can be either big or small, but the form factor of motherboard for which it is designed matters more. Laptops are computers that usually come in a clamshell form factor; however, in more recent years, deviations from this form factor, such as laptops that have a detachable screen that become tablet computers in their own right, have started to emerge.

Power supply unit

A power supply unit (PSU) converts alternating current (AC) electric power to low-voltage DC power for the internal components of the computer. Laptops are capable of running from a built-in battery, normally for a period of hours.

Motherboard

It is a board with integrated circuitry that connects the other parts of the computer including the CPU, the RAM, the disk drives (CD, DVD, hard disk, or any others) as well as any peripherals connected via the ports or the expansion slots.

★ ***New Words***

hardware ['hɑːdweə] n. 计算机硬件
collection [kə'lekʃ(ə)n] n. 集合
component [kəm'pəʊnənt] n. 部件；组件
constitute ['kɒnstɪtjuːt] vt. 组成，构成
system ['sɪstəm] n. 制度，体制；系统；方法
physical ['fɪzɪk(ə)l] adj. ［物］物理的；身体的；物质的
parts [pɑːts] n. ［机］零件，部件；形式
monitor ['mɒnɪtə] n. 监视器；显示屏；班长
keyboard ['kiːbɔːd] n. 键盘
data ['deɪtə] n. 数据(datum 的复数)；资料
storage ['stɔːrɪdʒ] n. 存储；仓库；贮藏所
graphic ['græfɪk] adj. 形象的；图表的；绘画似的
motherboard ['mʌðɚbɔrd] n. 底板；母板；主板
tangible ['tæn(d)ʒɪb(ə)l] adj. 有形的；可触摸的 n. 有形资产
object ['ɒbdʒɪkt] n. 目标；物体；宾语 vi. 反对
contrast ['kɒntrɑːst] vi. 对比；形成对照 vt. 使对比 n. 对比
software ['sɒf(t)weə] n. 软件
instruction [ɪn'strʌkʃənz] n. 指令；说明

store　[stɔː] n. 商店；储备　vt. 贮藏，储存
direct　[də'rekt] adj. 直接的；恰好的　vt. 导演；指向　vi. 指导；指挥
execute　['eksɪkjuːt] vt. 实行；执行
command　[kə'mɑːnd] vi. 命令，指挥　vt. 命令，控制　n. 控制；命令
combination　[kɒmbɪ'neɪʃ(ə)n] n. 结合；组合 [化学] 化合
form　[fɔːm] n. 形式，形状　vt. 构成，组成；排列
usable　['juːzəb(ə)l] adj. 可用的；合用的
personal　['pɜːs(ə)n(ə)l] adj. 个人的；亲自的　n. 人称代名词
common　['kɒmən] adj. 共同的；普通的；一般的；通常的　n. 普通
versatility　[ˌvɜːsə'tɪlətiː] n. 多功能性；多才多艺；用途广泛
relatively　['relətɪvlɪ] adv. 相当地；相对地，比较地
laptop　['læptɒp] n. 膝上型轻便电脑，笔记本电脑
generally　['dʒen(ə)rəlɪ] adv. 通常；普遍地，一般地
similar　['sɪmɪlə] adj. 相似的　n. 类似物
performance　[pə'fɔːm(ə)ns] n. 性能；绩效；表演；执行；表现
plastic　['plæstɪk] adj. 塑料的　n. 塑料制品
metal　['met(ə)l] n. 金属；合金　vt. 以金属覆盖　adj. 金属制的
enclosure　[ɪn'kləʊʒə] n. 附件；围墙；围场
house　[haʊs] vt. 覆盖 封装 给……房子住　n. 住宅；家庭；机构
desktop　['desktɒp] n. 桌面；台式机
fit　[fɪt] vt. 安装；使……适应
compact　[kəm'pækt] n. 合同　adj. 紧凑的，紧密的；简洁的　vt. 使简洁
design　[dɪ'zaɪn] vt. 设计；计划　n. 设计；图案
commonplace　['kɒmənpleɪs] n. 老生常谈；司空见惯的事　adj. 平凡的
all-in-one　['ɔː lin'wʌn] adj. 一件式的；连身的　n. 一体化
style　[staɪl] n. 风格；时尚；
iMac abbr. Integrated Microwave Amplifier Converter 苹果一体机
factor　['fæktə] n. 因素；要素
clamshell　['klæmʃel] n. 掀盖式
deviation　[diːvɪ'eɪʃ(ə)n] n. 偏差；误差；背离
detachable　[dɪ'tætʃəbl] adj. 可分开的；可拆开的
tablet　['tæblɪt] n. 平板电脑
emerge　[ɪ'mɜːdʒ] vi. 浮现；摆脱；暴露
supply　[sə'plaɪ] n. 供给，补给　vt. 供给
convert　[kən'vɜːt] vt. 使转变；转换
alternating　['ɔːltəneɪtɪŋ] adj. 交替的；交互的
current　['kʌr(ə)nt] adj. 现在的　n. (电)流
low-voltage　['ləu'vəultidʒ] adj. 低压的

internal [ɪn'tɜːn(ə)l] n. 本质 adj. 内部的
capable ['keɪpəb(ə)l] adj. 能干的，能胜任的
built-in [ˌbɪlt 'ɪn] adj. 嵌入的；固定的
battery ['bætri] n. [电] 电池
normally ['nɔːm(ə)lɪ] adv. 正常地；通常地
integrated ['ɪntɪgreɪtɪd] adj. 综合的；完整的；互相协调的
circuitry ['sɜːkɪtrɪ] n. 电路；电路系统
peripheral [pə'rɪf(ə)r(ə)l] adj. 外围的；次要的 n. 外部设备
via ['vaɪə] prep. 渠道，通过；经由
port [pɔːt] n.（计算机的）端口
expansion [ɪk'spænʃ(ə)n] n. 膨胀
slot [slɒt] n. 位置；插槽

★ ***Phrases and Expressions***

physical part	物理部件
such as	诸如，例如
due to	由于
computer case	电脑机箱
either... or	或者……或者
power supply unit	电源（箱）
clamshell form factor	翻盖形式
alternating current(AC)electric power	交流电
DC power	直流电
expansion slot	扩充插槽

Task Ⅲ－1

1. Fill in the blanks without referring to the passage.

Computer hardware is the ________(1) of physical components that constitute a computer ________(2). Computer hardware is the physical parts or ________(3) of a computer, such as monitor, ________(4), computer data storage, graphic card, sound card, motherboard, and so on, all of which are tangible objects. By contrast, ________(5) is instructions that can be ________(6) and run by hardware.

2. Answer the following questions according to the passage.

(1) What is the computer hardware?

____

(2) What is a personal computer?

____

(3) What role does a power supply unit play in a personal computer?

____

(4) How do laptops work?

____

(5) What role does the motherboard of a personal computer play?

____

3. Complete each of the following statements according to the passage.

(1) Hardware is directed by ________ to execute any command or instruction.

(2) A usable computing system is formed by ________ of hardware and software.

(3) A computer case can be either big or small, but ________ of motherboard for which it is designed matters more.

(4) Laptops have a detachable screen that become ________ in their own right, and they have started to emerge.

(5) Laptops are capable of running from ________.

(6) The computer motherboard is a board with integrated circuitry that connects ________ of the computer.

4. Translate the following sentences into English.

(1) 我把我的一些东西存放起来。(in storage)

____

(2) 看看他们的新系统，我们的相比之下就显得太过时了。(by contrast)

____

(3) 彼得决定直接和经理联系。(directly)

____

(4) 我喜欢诸如茶和汽水之类的饮料。(such as)

____

(5) 这家公司既经营硬件，又经营软件。(hardware, software)

____

(6) 这些岩石是40多亿年前形成的。(form)

____

(7) 个人电脑确实推动了一个科学和商业的新时代的到来。(personal computer)

____

(8) 你喝茶也行，喝咖啡也行。(either... or)

____

5. Translate the following sentences into Chinese.

(1) Computer hardware is the physical parts or components of a computer, such as monitor, keyboard, computer data storage, graphic card, sound card, motherboard, and so on, all of which are tangible objects.

____

____

(2) Hardware is directed by the software to execute any command or instruction.

____

(3) A personal computer, also known as the PC, is one of the most common types of computer due to its versatility and relatively low price.

____

(4) The computer case is a plastic or metal enclosure that houses most of the components.

____

(5) However, in recent years more compact designs have become more commonplace, such as the all-in-one style designs from Apple, namely the iMac.

____

____

(6) Laptops are computers that usually come in a clamshell form factor; however, in more recent years, deviations from this form factor, such as laptops that have a detachable screen that become tablet computers in their own right, have started to emerge.

____

____

(7) A power supply unit (PSU) converts alternating current (AC) electric power to low-voltage DC power for the internal components of the computer.

____

____

(8) Laptops are capable of running from a built—in battery, normally for a period of hours.

____

(9) The computer motherboard is the main component of a computer.

____

(10) The computer motherboard connects the other parts of the computer including the CPU, the RAM, the disk drives (CD, DVD, hard disk, or any others) as well as any peripherals connected via the ports or the expansion slots.

____

____

Passage B

Computer Hardware(Ⅱ)

Expansion card

An expansion card in computing is a printed circuit board that can be inserted into an expansion slot of a computer motherboard or backplane to add functionality to a computer system via the expansion bus. Expansion cards can be used to obtain or expand on features not offered by the motherboard.

Computer data storage

A storage device is any computing hardware and digital media that is used for storing, porting and extracting data files and objects. It can hold and store information both temporarily and permanently, and can be internal or external to a computer, server or any similar computing device. Data storage is a core function and fundamental component of computers.

Fixed media

Data is stored by a computer using a variety of media. Hard disk drives are found in virtually all older computers, due to their high capacity and low cost, but solid-state drives are faster and more power efficient, although currently more expensive than hard drives in terms of dollar per gigabyte, so are often found in personal computers built post-2007. Some systems may use a disk array controller for greater performance or reliability.

Removable media

To transfer data between computers, a USB flash drive or optical disc may be used. Their usefulness depends on being readable by other systems; the majority of machines have an optical disk drive, and virtually all have at least one USB port.

Input and output peripherals

Peripheral

Input and output devices are typically housed externally to the main computer chassis. The following are either standard or very common to many computer systems.

Input device

Input devices allow the user to enter information into the system, or control its operation. Most personal computers have a mouse and keyboard, but laptop systems typically use a touchpad instead of a mouse. Other input devices include webcams, microphones, joysticks, and image scanners.

Output device

Output devices display information in a human readable form. Such devices could include printers, speakers, monitors or a Braille embosser.

★ ***New Words***

computing [kəmˈpjuːtɪŋ] n. 计算；处理；从事电脑工作

circuit [ˈsɜːkɪt] n. 电路；线路；环形；回路 v. 巡回

insert [ɪnˈsɜːt] vt. 插入；嵌入 n. 插入物；管芯

backplane [ˈbæk ˌplein] n. 背板，[电子] 底板；基架

functionality [fʌŋkʃəˈnælətɪ] n. 功能；[数] 泛函性，函数性

obtain [əbˈteɪn] vi. 获得；流行 vt. 获得

expand [ɪkˈspænd; ek-] vt. 扩张；使膨胀；详述 vi. 发展；张开，展开

feature [ˈfiːtʃə] n. 特征；容貌 vi. 起重要作用 vt. 特写；以……为特色
offer [ˈɒfə] vt. 提供；试图 n. 提议；出价；录取通知书 vi. 出现
device [dɪˈvaɪs] n. 装置；策略；图案
digital [ˈdɪdʒɪt(ə)l] adj. 数字的；手指的 n. 数字；键
media [ˈmiːdɪə] n. 媒体；媒质(medium 的复数)
porting [pɔrtɪŋ] n. 移植，出口商品 v. 携带；搬运(port 的现在分词)
extract [ˈekstrækt] vt. 提取；取出；摘录；榨取 n. 汁；摘录
file [faɪl] n. 文件；档案；文件夹 vt. 提出；琢磨；把……归档
information [ɪnfəˈmeɪʃ(ə)n] n. 信息，资料；知识；情报；通知
temporarily [ˈtemp(ə)r(ər)ɪlɪ] adv. 临时地，临时
permanently [ˈpɜːm(ə)nəntlɪ] adv. 永久地，长期不变地
internal [ɪnˈtɜːn(ə)l] n. 本质 adj. 内部的；体内的；(机构)内部的
external [ɪkˈstɜːn(ə)l; ek-] adj. 外部的；外国的 n. 外部；外面
server [ˈsɜːvə] n. (计算机)主机，发球员；服伺者
core [kɔː] n. 核心；要点；果心；[计] 磁心
function [ˈfʌŋ(k)ʃ(ə)n] n. 功能；[数] 函数；职责 vi. 运行
fundamental [fʌndəˈment(ə)l] adj. 基本的，根本的 n. 基本原理
variety [vəˈraɪətɪ] n. 多样；种类；杂耍；变化，多样化
virtually [ˈvəːtʃʊəli] adv. 事实上，几乎；实质上
solid-state [ˈsɒlɪdˌsteɪt] adj. 固态的；固态电子学的；使用电晶体的
efficient [ɪˈfɪʃ(ə)nt] adj. 有效率的；有能力的；生效的
currently [ˈkʌrəntlɪ] adv. 当前；一般地
gigabyte [ˈgɪgəbaɪt] n. 十亿字节；十亿位组
post- [pəust-] pref. 表示"后、在……之后"的意思
array [əˈreɪ] n. 数组；排列，列阵；大批，一系列 vt. 排列
controller [kənˈtrəʊlə] n. 控制器；管理员
reliability [rɪˌlaɪəˈbɪlətɪ] n. 可靠性
removable [rɪˈmuːvəbl] adj. 可移动的；可去掉的；可免职的
transfer [trænsˈfɜ] n. 转移；传递；过户 vi. 转让；转学 vt. 使转移
USB abbr. 端口，通用串行总线(Universal Serial Bus)
flash [flæʃ] n. 闪光；交光灯；瞬间场面，瞬间镜头
optical [ˈɒptɪk(ə)l] adj. 光学的；眼睛的，视觉的
disc [dɪsk] n. 圆盘，[电子] 唱片(等于 disk)
usefulness [ˈjuːsf(ʊ)lnəs] n. 有用；有效性；有益
readable [ˈriːdəb(ə)l] adj. 可读的；易读的
majority [məˈdʒɒrɪtɪ] n. 多数；成年
machine [məˈʃiːn] n. 机械，机器；机械般工作的人 vt. 用机器制造
typically [ˈtɪpɪkəlɪ] adv. 代表性地；作为特色地
externally [ɪkˈstəːnli] adv. 外部地；外表上，外形上

chassis [ˈʃæsɪ] n. 底盘，底架
standard [ˈstændəd] n. 标准；水准 adj. 标准的
allow [əˈlaʊ] vt. 允许；给予；认可 vi. 容许；考虑
enter [ˈentə] n. [计] 输入；回车 vt. 进入；开始；参加 vi. 进去
control [kənˈtrəʊl] n. 控制；管理；抑制
operation [ɒpəˈreɪʃ(ə)n] n. 操作；经营；[外科] 手术；[计] 运算
touchpad [ˈtʌtʃpæd] n. 触摸屏设备，触摸板
webcam [ˈwebkæm] n. 网络摄像头
microphone [ˈmaɪkrəfəʊn] n. 扩音器，麦克风
joystick [ˈdʒɔɪstɪk] n.（计算机游戏的）操纵杆，摇杆，控制杆
image [ˈɪmɪdʒ] n. 影像；想象
scanner [ˈskænə] n.（计）扫描仪；扫描器
display [dɪˈspleɪ] n. 显示；炫耀 vt. 显示；表现；陈列
human [ˈhjuːmən] adj. 人的；人类的 n. 人；人类
printer [ˈprɪntə] n. [计] 打印机；印刷工
speaker [ˈspiːkə] n. 扬声器；说话者；演讲者
embosser [imˈbɔsə] n. [皮革] 压纹机；[皮革] 压花机

★ ***Phrases and Expressions***

expansion card	扩充插件板
printed circuit board	印刷电路板
expansion bus	扩展总线
computer data storage	计算机数据存储器
fixed media	虚拟硬盘
hard disk drive	硬盘驱动器(亦作 hard drive)
in terms of	依据；按照；在……方面
removable media	可移动媒介
USB flash drive	闪存盘，随身碟，优盘
optical disc	(计)光盘
depends on	依赖于
at least	至少
image scanner	(计)图像扫描仪
Braille embosser	盲人点字浮雕器

Task Ⅲ - 2

1. Match the following English phrases in Column A with their Chinese equivalents in Column B.

A	B
(1) expansion slot	a. 物理部件

(2) physical part b. 图像扫描仪
(3) alternating current electric power c. 虚拟硬盘
(4) clamshell form factor d. 扩充插槽
(5) power supply unit e. 硬盘驱动器
(6) computer data storage f. 翻盖形式
(7) hard disk drive g. 交流电
(8) optical disc h. 计算机数据存储器
(9) image scanner i. 电源(箱)
(10) fixed media j. 光盘

2. Decide whether the following statements are T(true) or F (false) according to the passage.

(1) A computer expansion card in computing is a circuit board. ( )

(2) Computer expansion cards can be used to obtain or expand on features offered by the computer motherboard. ( )

(3) A computer storage device is any computing hardware and digital media that is used for storing, porting and extracting data files and objects. ( )

(4) A storage device can hold and store information both temporarily and permanently. ( )

(5) Data storage in a computer is a common function and fundamental component of computers. ( )

(6) Data is stored by a computer using a variety of media. ( )

(7) Some computer systems may use a disk array controller for greater performance or reliability. ( )

(8) To transfer data between computers, only an optical disc may be used in a personal computer. ( )

(9) Input and output devices in a computer are typically housed externally to the main computer chassis. ( )

(10) Output devices in a computer could include printers, speakers, monitors or a Braille embosser. ( )

3. Translate the following passage into Chinese.

Data is stored by a computer using a variety of media. Hard disk drives are found in virtually all older computers, due to their high capacity and low cost, but solid-state drives are faster and more power efficient, although currently more expensive than hard drives in terms of dollar per gigabyte, so are often found in personal computers built post-2007. Some systems may use a disk array controller for greater performance or reliability.

Read the following passage and choose the best answer to each question.

Computer Motherboard

The computer motherboard is the main component of a computer. Components directly attached to or to part of the motherboard include:

The CPU (Central Processing Unit), which performs most of the calculations which enable a computer to function, and is sometimes referred to as the brain of the computer. It is usually cooled by a heat sink and fan, or water-cooling system. Most newer CPUs include an on-die Graphics Processing Unit (GPU). The clock speed of CPUs governs how fast it executes instructions, and is measured in GHz; typical values lie between 1 GHz and 5 GHz. Many modern computers have the option to over clock the CPU which enhances performance at the expense of greater thermal output and thus a need for improved cooling.

The chipset, which includes the north bridge, mediates communication between the CPU and the other components of the system, including main memory.

Random-Access Memory (RAM), which stores the code and data that are being actively accessed by the CPU. For example, when a web browser is opened on the computer it takes up memory; this is stored in the RAM until the web browser is closed. RAM usually comes on DIMMs in the sizes 2GB, 4GB, and 8GB, but can be much larger.

Read-Only Memory (ROM), which stores the BIOS that runs when the computer is powered on or otherwise begins execution, a process known as Bootstrapping, or "booting" or "booting up". The BIOS (Basic Input Output System) includes boot firmware and power management firmware. Newer motherboards use Unified Extensible Firmware Interface (UEFI) instead of BIOS.

Buses that connect the CPU to various internal components and to expand cards for graphics and sound.

The CMOS battery, which powers the memory for date and time in the BIOS chip. This battery is generally a watch battery.

The video card (also known as the graphics card), which processes computer graphics. More powerful graphics cards are better suited to handle strenuous tasks, such as playing intensive video games.

1. What is the Central Processing Unit?
   A. It is the computer motherboard of a computer.
   B. It performs most of the calculations which enable a computer to function, and is sometimes referred to as the brain of the computer.
   C. It is calculations which enable a computer to function, and is sometimes referred to as the brain of the computer.

D. It is always referred to as the brain of the computer.

2. How is the CPU (Central Processing Unit) usually cooled?

   A. It is usually cooled by an air conditioner.

   B. It is usually cooled by ice-cooling system.

   C. It is usually cooled by water recycling.

   D. It is usually cooled by a heat sink and fan, or water-cooling system.

3. What size does RAM usually come on DIMMs in?

   A. It usually comes on DIMMs in the sizes 2GB.

   B. It usually comes on DIMMs in the sizes 4GB.

   C. It usually comes on DIMMs in the sizes 8GB.

   D. All of the above.

4. What does the BIOS (Basic Input Output System) include?

   A. It includes boot firmware, power management firmware and buses.

   B. It includes boot firmware and buses.

   C. It includes boot firmware and power management firmware.

   D. It includes power management firmware and buses.

5. What does a computer motherboard mainly include according the passage?

   A. It includes CPU, the chipset, RAM, ROM, buses, CMOS battery, video card.

   B. It includes CPU, the chipset, RAM, ROM, buses, CMOS battery.

   C. It includes CPU, the chipset, RAM, ROM, buses, video card.

   D. It includes CPU, ROM, buses, CMOS battery, video card.

Section Four Writing

电话留言条 & 名片(Telephone Message & Name Card)

1. 电话留言条(Telephone Message)

电话留言条是某人打来电话时，所找的人不在场，你替他或她接了电话后留给该人的字条。下面是电话留言条常见的格式：

To: ________(收留言条者)

Time: ________(来电时间)

Message: ____

____.(来电内容)

From: ________(电话记录者)

◆常用语句◆

① Mr. Smith's appointment with you tomorrow has to be cancelled.（史密斯先生不得不取消明天与你的约会。）

② She will contact you for another appointment when she is back.（当她回来时会再与你约时间会面。）

2. 名片(Name Card)

在社交和公务活动中，人们初次见面时应互赠名片，它是日后联系的线索。名片的形式多样，基本内容包括姓名、职务、所在单位及联系方式。下面是名片常见的格式。

________________________(单位名称)	
______(姓名)	
______(职务/职称)	
__________(地址)	__________(电话)
__________(邮编)	__________(传真、邮箱地址等)

◆常用词语◆

① 职务/职称

president（校长）；dean（主任）；professor（教授）；associate professor（副教授）；director（董事长）；head of the bureau（局长）；general manager（总经理）；assistant manager（经理助理）；personnel manager/director（人力资源部主任）；chief engineer（总工程师）；senior engineer（高级工程师）；accountant（会计师）；official（公务员）

② 单位名称

CO. LTD（limited company）(有限公司)；Corp.（corporation）（集团）；group company（集团公司）；bureau（局）；department(系/部)；section(处/科)；agency(社)；institute(所)；office(办公室)

③ 联系方式

Add.（address)(地址)；P. C.（post code)(邮编)；Tel.(telephone)(电话)；O.(办公室)；H.(家/宅)；EM.(E—mail)(邮箱地址)；fax(传真)；mobile phone(手机)

Task Ⅳ

1. You are required to write a Telephone Message according to the following instructions given in Chinese.

说明：按电话留言的格式和要求，以秘书 Jane 的名义，给 Mr. William 写一份电话留言，包括以下内容。

(1) 来电人：Honeywell(霍尼韦尔公司)公司的 Mr. Black；

(2) 来电时间：2016 年 6 月 25 日上午 10 时；

(3) 事由：Mr. Black 明天要出差去北京，因此原定与 Mr. William 后天上午的约会只能取消，等出差回来后再与 Mr. William 约时间会面。

2. You are required to make a Name Card according to the following instructions given in Chinese.

我的名字叫王佳伟。我工作在沈阳国际语言学校外事处。

其它个人信息提供如下：

职务：主任（教授）　地址：沈阳市皇姑区崇山路 666 号

电话 & 传真：024-22235555　　　E-mail 地址：gaop702@163.com

邮编：110102

Unit Two Computer Software

Section One Speaking

Task I

1. Look at the following graphic and speak out each part of Software Development Tools in Chinese.

Software Development Tools

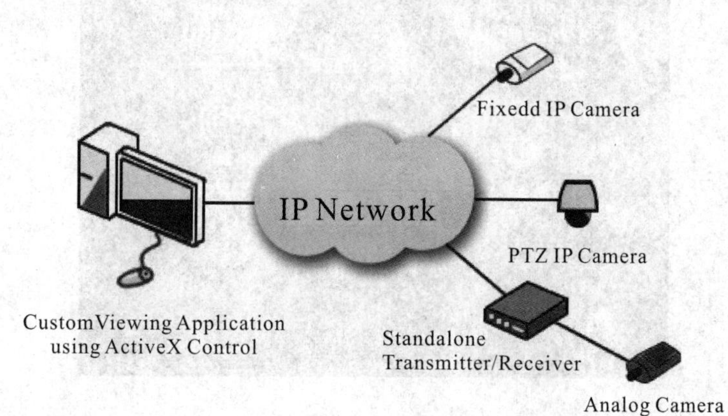

IP Network ________ Custom Viewing Application ________

Fixed IP Camera ________ PTZ IP Camera ________

Analog Camera ________ Standalone Transmitter\Receiver ________

2. List the Chinese names of the terms in the following graphic.

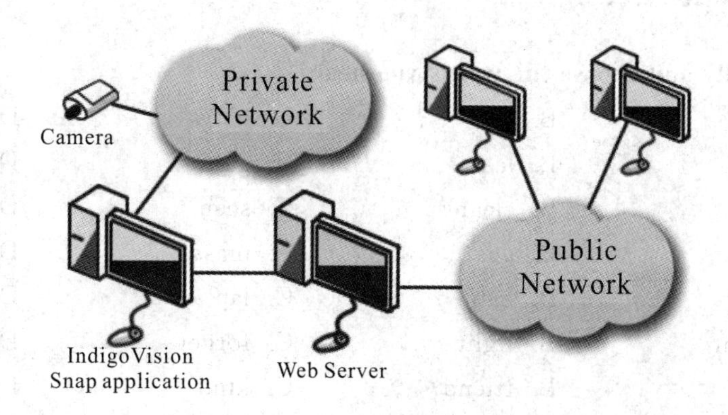

3. Suppose you are a computer engineer. Give a presentation to each picture in English.

Picture 1

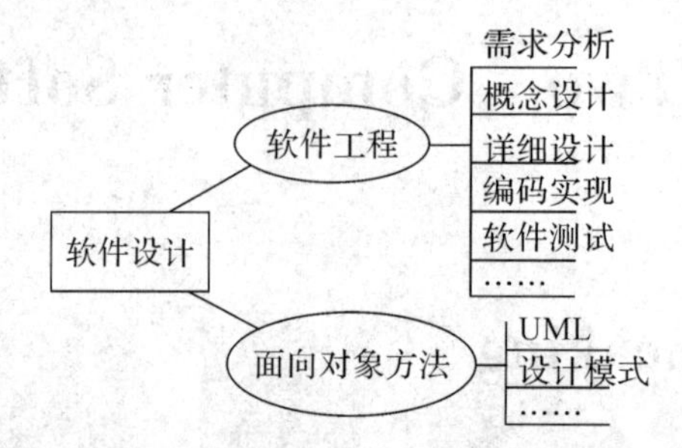

Picture 2

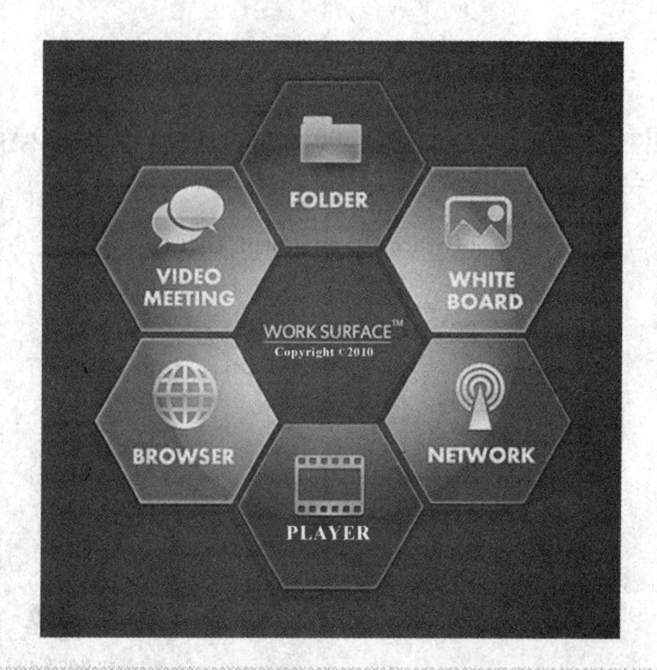

Section Two Listening

Task Ⅱ

1. Listen carefully and choose the words you hear.

(1) A. now B. cow C. blow D. how

(2) A. bake B. lake C. gate D. get

(3) A. keen B. clean C. seen D. lean

(4) A. has B. gas C. mass D. vase

(5) A. fat B. map C. lap D. nap

(6) A. flight B. fight C. forget D. fat

(7) A. find B. friend C. kind D. fond

(8) A. dot B. got C. not D. hot
(9) A. lone B. home C. tone D. bone
(10) A. daughter B. doctor C. taught D. dot

2. Listen to the short passage and choose the proper words to fill in the blanks.

Linda is my best friend. She is 15 years old. She is a __1__ (nice, pretty, beautiful) girl with a round face and two big black eyes. She __2__ (always, often, sometimes) has a smile on her face. She is taller than I.

Every morning we go to __3__ (park, playground, school) together. She studies quite well and she's a top student in our class. She is modest in her behavior. When I have difficulty in __4__ (English, Chinese, Chemistry), I always ask her for help. We are both interested in __5__ (sports, music, reading). At weekends, we join the same hobby group and play the violin together. We like each other.

3. Listen to the dialogues and choose the best responses to what you hear.

(1) A. I think it flew at 48 kilometers per hour.
B. I think it flew 36.5 meters far.
C. I think it flew 3.5 meters high.
D. I think it flew 112 seconds. ______

(2) A. It's Monday.
B. It's windy.
C. I like it.
D. I have enough clothes. ______

(3) A. Thank you.
B. Don't thank me.
C. That's right.
D. You are welcome. ______

(4) A. Yes, I do.
B. Yes, I did.
C. Yes, I have.
D. Yes, I am. ______

(5) A. No, I haven't.
B. Yes, I do.
C. Oh, me? Apple pie.
D. No, I don't. ______

4. Listen to the dialogues and choose the best answer to each question you hear.

(1) A. He works there as a Chinese teacher. B. He is worker.
C. He is a farmer. D. He is a driver.
(2) A. He didn't like the trip. B. He didn't like the weather.
C. He didn't like the food. D. He didn't like the people.
(3) A. Teacher and student. B. Father and daughter.

C. Friends. D. Boss and clerk.

(4) A. At the airport. B. In a shop.

C. In a club. D. In an office.

(5) A. A doctor. B. A teacher.

C. A bank clerk. D. A nurse.

5. Listen to the dialogue and answer the following questions by filling in the blanks.

(1) Who is Wang Hai talking with?

She is ____________.

(2) How has Li Hua changed?

She has changed ____________.

(3) Li Hua has long hair or short hair now?

She has ____________ now.

(4) Now Li Hua's eyes are shortsighted or not?

Now her eyes ____________.

(5) Who used to play the piano in the dialogue?

____________ used to play the piano.

6. Listen to the passage and fill in the blanks with words you hear. Some new words given below will be of some help to you.

cash [kæʃ] n. 现金

register ['redʒɪstə] v. 登记；注册

order ['ɔːdə] v. 点菜，预订

divide [dɪ'vaɪd] v. 分成

pile [paɪl] n. 堆

neatly ['niːtlɪ] adj. 整洁地，干净地

sip [sɪp] v. 小口喝

refuse [rɪ'fjuːz] v. 拒绝

politely [pə'laɪtlɪ] adv. 有礼貌地

cash register 收银处

French fries 炸薯条

in front of 在……前面

take turns 依次；轮流

come over 过来

Old Couple at McDonald's

A little old man and his wife walked slowly into McDonald's on cold __1__ evening. They __2__ a table near the back wall, and then the little old man walked to the cash register to order. After a while he got the food back and they began to open it.

There was one hamburger, some French fries and one drink. The little old man carefully cut the hamburger in half and divided the French fries in two piles. Then he neatly __3__ the half of the food in front of his wife. He took a sip of the drink and his wife took a sip. "How poor the old people are!" the people __4__ them thought. As the man began to __5__ his hamburger and his French fries, his wife sat there watching him and took turns to __6__. A young man came over and offered to __7__ another meal for them. __8__ they refused politely and said that they got used to sharing everything.

Then a young __9__ asked a question to the little old lady. "Madam, why aren't you eating? You said that you __10__ everything. Then what are you waiting for?" She answered, "The teeth."

Section Three Reading

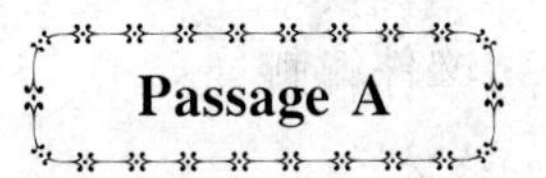

Computer Software

Computer software, or simply software, is a part of a computer system that consists of data or computer instructions, in contrast to the physical hardware from which the system is built. In computer science and software engineering computer software is all information processed by computer systems, programs and data. Computer software includes computer programs, libraries and related non-executable data, such as online documentation or digital media. Computer hardware and software require each other and neither can be realistically used on its own.

At the lowest level, executable code consists of machine language instructions specific to an individual processor—typically a central processing unit (CPU). A machine language consists of groups of binary values signifying processor instructions that change the state of the computer from its preceding state. For example, an instruction may change the value stored in a particular storage location in the computer — an effect that is not directly observable to the user. An instruction may also (indirectly) cause something to appear on a display of the computer system—a state change which should be visible to the user. The processor carries out the instructions in the order they are provided, unless it is instructed to "jump" to a different instruction, or is interrupted (by now multi-core processors are dominant, where each core can run instructions in order; then, however, each application software runs only on one core by default, but some software has been made to run on many).

The majority of software is written in high-level programming languages that are easier and more efficient for programmers to use because they are closer than machine languages to natural languages. High-level programming languages are translated into machine language using a compiler or an interpreter or a combination of the two. Software may also be written in a low-level assembly language, which has strong correspondence to the computer's machine language instructions and is translated into machine language using an assembler.

★ ***New Words***

simply [ˈsɪmplɪ] adv. 简单地；仅仅；简直；朴素地
engineering [endʒɪˈnɪərɪŋ] n. 工程；工程学 v. 设计
process [prəˈses] vt. 处理；加工
program [ˈprəʊgræm] n. 程序；计划 vt. 为……制订计划 vi. 编程序
library [ˈlaɪbrərɪ] n. 图书馆；藏书室；文库
relate [rɪˈleɪt] vt. 叙述；使……有联系 vi. 涉及；认同；符合
non-executable [ˌnʌnɪgˈzekjʊtəb(ə)l] adj. 不可执行的；不可实行的
online [ɒnˈlaɪn] adj. 联机的；在线的 adv. 在线地
documentation [ˌdɒkjʊmenˈteɪʃ(ə)n] n. 文件；证明文件；史实；文件编制
require [rɪˈkwaɪə] vt. 需要；要求；命令
realistically [ˌriəˈlɪstɪkli] adv. 现实地；实际地
executable [ɪgˈzekjʊtəb(ə)l] adj. 可执行的；可实行的
code [kəʊd] n. 代码，密码；编码；法典 vt. 编码
language [ˈlæŋgwɪdʒ] n. 语言；语言文字；表达能力
specific [spəˈsɪfɪk] adj. 特殊的，特定的；明确的；详细的 n. 特性；细节
individual [ˌɪndɪˈvɪdʒʊəl] adj. 个人的；个别的；独特的 n. 个人，个体
processor [ˈprəʊsesə] n. [计] 处理器；处理程序；加工者
binary [ˈbaɪnərɪ] adj. [数] 二进制的；二元的，二态的
value [ˈvæljuː] n. 值；价值；价格；重要性；确切含义
signify [ˈsɪgnɪfaɪ] n. 代表；预示 v. 象征；预示
change [tʃeɪn(d)ʒ] vt. 改变；交换 n. 变化
state [steɪt] n. 国家；州；情形 vt. 规定；声明；陈述
preceding [prɪˈsiːdɪŋ] adj. 在前的；前述的 v. 在……之前
particular [pəˈtɪkjʊlə(r)] adj. 特别的；详细的；独有的 n. 详细说明
location [lə(ʊ)ˈkeɪʃ(ə)n] n. 位置；地点
effect [ɪˈfekt] n. 影响；效果；作用 vt. 产生；达到目的
observable [əbˈzɜːvəbl] adj. 觉察得到的；看得见的 n.感觉到的事物
cause [kɔːz] n. 原因；事业；目标 vt. 引起
appear [əˈpɪə] vi. 出现；显得；似乎；出庭；登场
visible [ˈvɪzəbl] adj. 明显的；看得见的；现有的；可得到的 n. 可见物
provide [prəˈvaɪd] vt. 提供；规定；准备；装备 vi. 规定；抚养
interrupt [ɪntəˈrʌpt] vt. 中断；打断；插嘴；妨碍
multi-core [ˈmʌtɪkɒ] n. 多核，多核心
dominant [ˈdɒmɪnənt] adj. 显性的；占优势的；支配的，统治的 n. 显性

application [ˌæplɪˈkeɪʃ(ə)n] n. 应用；申请；应用程序
default [dɪˈfɔːlt; ˈdiːfɔːlt] vi. 拖欠；不履行 n. 违约，系统默认值 vt. 不履行
programmer [ˈprəʊgræmə] n.（计）程序设计员
natural [ˈnætʃrəl] adj. 自然的；物质的；天生的；不做作的 n. 自然的事情
high-level [ˌhaɪˈlevl] adj. 高级的；高阶层的；在高空的
compiler [kəmˈpaɪlə] n. 编译器；[计] 编译程序；编辑者，汇编者
interpreter [ɪnˈtɜːprɪtə] n. 解释者；口译者；注释器
assembly [əˈsemblɪ] n. 装配；集会，集合
correspondence [kɒrɪˈspɒnd(ə)ns] n. 通信；一致；相当
translate [trænsˈleɪt] vt. 翻译；转化；解释；转变为；调动 vi. 翻译
assembler [əˈsemblə] n. 汇编程序；汇编机；装配工

★ ***Phrases and Expressions***

consists of	由……构成；包括
in contrast to	与……对比(或对照)
each other	互相，彼此
on one's own	依靠自己
binary value	二进制数值
carry out	执行，实行；贯彻；实现
application software	应用软件
assembly language	（计）汇编语言

Task Ⅲ－1

1. Fill in the blanks without referring to the passage.

Computer software, or simply software, is a part of a computer system that consists of data or ________(1), in contrast to the physical hardware from which the system is built. In computer science and software engineering computer software is all ________(2) processed by computer systems, programs and ________(3). Computer software includes computer programs, ________(4) and related non－executable data, such as online documentation or ________(5). Computer hardware and software require ________(6) and neither can be realistically used on its own.

2. Answer the following questions according to the passage.

(1) What is computer software?

____

(2) Which parts does computer software include?

____

(3) What role does a machine language play in a computer?

____

(4) How does the processor carry out the instructions in a computer?

____

(5) Why is the majority of software written in high-level programming languages?

____

3. Complete each of the following statements according to the passage.

(1) Computer software is a part of a computer system that consists of ________ or computer instructions.

(2) Computer hardware and software require each other and neither can ________ on its own.

(3) An processor instruction may change ________ stored in a particular storage location in the computer.

(4) Each application software runs only on one core by ________, but some software has been made to run on many.

(5) The majority of software is written in high－level programming languages because they are closer than machine languages to ________.

(6) High-level programming languages are translated into machine language using ________ or an interpreter or a combination of the two.

4. Translate the following sentences into English.

(1) 他们很熟悉这一观点，所有的物质都是由原子构成的。(consist of)

____

(2) 我可以帮助你学习这个新电脑程序。(program)

____

(3) 他与以前相比，显得更健康些。(in contrast to)

____

(4) 源代码的此位置没有可执行代码。(executable code)

____

(5) 我们需要实行我们的计划。(carry out)

____

(6) 应用软件为用户完成具体的任务。(application software)

____

(7) 我们就要用汇编语言，甚至要用机器语言去编写程序。(assembly language)

____

(8) 这东西看不出有明显的用途。(visible)

____

5. Translate the following sentences into Chinese.

(1) In computer science and software engineering computer software is all information processed by computer systems, programs and data.

____

____

(2) Computer software includes computer programs, libraries and related non-executable data, such as online documentation or digital media.

____

(3) A machine language consists of groups of binary values signifying processor instructions that change the state of the computer from its preceding state.

____

(4) An instruction may also (indirectly) cause something to appear on a display of the computer system—a state change which should be visible to the user.

____

(5) The processor carries out the instructions in the order they are provided, unless it is instructed to "jump" to a different instruction, or is interrupted.

____

(6) By now multi-core processors are dominant, where each core can run instructions in order.

____

(7) The majority of software is written in high-level programming languages that are easier and more efficient for programmers to use.

____

(8) High-level languages are translated into machine language using a compiler or an interpreter or a combination of the two.

____

(9) Software may also be written in a low-level assembly language, which has strong correspondence to the computer's machine language instructions and is translated into machine language using an assembler.

____

____

Passage B

Application Software and System Software

On virtually all computer platforms software can be grouped into a few broad categories. Based on the goal of purpose, or domain of its use, computer software can be divided into application software and system software.

Application software

Application software is software that uses the computer system to perform special functions or provide entertainment functions beyond the basic operation of the computer itself. There are many different types of application software because the range of tasks that can be performed with a modern computer is so large.

System software

System software is software that directly operates the computer hardware, to provide basic functionality needed by users and other software, and to provide a platform for running application software. System software includes:

Operating systems

Operating systems are essential collections of software that manage resources and provides common services for other software that runs "on top" of them. Supervisory programs, boot loaders, shells and window systems are core parts of operating systems. In practice, an operating system comes bundled with additional software (including application software) so that a user can potentially do some work with a computer that only has one operating system.

Device drivers

Device drivers operate or control a particular type of device that is attached to a computer. Each device needs at least one corresponding device driver; because a computer typically has at minimum at least one input device and at least one output device, a computer typically needs more than one device driver.

Utilities

Utilities are computer programs designed to assist users in the maintenance and care of their computers.

PS: Malicious software or malware

Malicious software or malware is software that is developed to harm and disrupt computers. As such, malware is undesirable. Malware is closely associated with computer-related crimes, though some malicious programs may have been designed as practical jokes.

★ ***New Words***

group [gru:p] n. 组；分(类) adj. 群的；团体的 vi. 聚合 把……分组

broad [brɔ:d] adj. 宽的，辽阔的 n. 宽阔部分 adv. 宽阔地

category ['kætɪg(ə)rɪ] n. 种类，分类；[数] 范畴

goal [gəʊl] n. 目标；球门，得分数；终点

purpose ['pɜ:pəs] n. 目的；用途；意志 vt. 决心

domain [də(ʊ)'meɪn] n. 领域；域名；产业；地产

perform [pə'fɔ:m] vt. 执行；完成；演奏 vi. 执行，机器运转

special ['speʃ(ə)l] n. 特使；特刊；专车；特价商品 adj. 特别的

beyond [bɪ'jɒnd] prep. 超过；越过；在……较远的一边 adv. 在远处

range [reɪn(d)ʒ] n. 范围；幅度；排；山脉 vi. 平行；延伸 vt. 归类于

task [tɑ:sk] vt. 分派任务 n. 工作，作业；任务

essential [ɪ'senʃ(ə)l] adj. 基本的；必要的；本质的 n. 本质；要素

manage ['mænɪdʒ] vt. 管理；经营；控制；设法 vi. 处理

resource [rɪ'sɔ:s; rɪ'zɔ:s] n. 资源，财力；办法；智谋

supervisory ['sju:pəˌvaɪzərɪ] adj. 监督的

shell [ʃel] n. 壳(命令解析器)
practice ['præktɪs] n. 实践；练习；惯例 vi. 练习
bundle ['bʌnd(ə)l] n. 束；捆 vt. 捆 vi. 匆忙离开
additional [ə'dɪʃ(ə)n(ə)l] adj. 附加的，额外的
potentially [pə'tɛnʃəli] adv. 可能地，潜在地
attach [ə'tætʃ] vt. 使依附；贴上；系上；使依恋 vi. 附加
corresponding [ˌkɒrɪ'spɒndɪŋ] adj. 相当的；一致的；通信的 v. 类似
minimum ['mɪnɪməm] n. 最小值；最低限度；最小量 adj. 最小的
utilities [juː'tɪlɪtɪz] n. 公用事业；(计) 实用程序
assist [ə'sɪst] n. 帮助；助攻 vi. 参加；出席 vt. 帮助
maintenance ['meɪntənəns] n. 维护，维修；保持
malware [mælwɛə] n. 恶意软件
harm [hɑːm] n. 伤害；损害 vt. 伤害
disrupt [dɪs'rʌpt] vt. 破坏；使瓦解；使分裂；使中断；使陷于混乱
undesirable [ʌndɪ'zaɪərəb(ə)l] adj. 不良的；不受欢迎的 n. 不良分子
associate [ə'səʊʃɪeɪt] vi. 交往；结交 n. 同事，伙伴 vt. 联想
crime [kraɪm] n. 罪行，犯罪；罪恶；犯罪活动 vt. 控告……违反纪律

★ ***Phrases and Expressions***

system software	系统软件
based on	基于
divide into	把……分成
operating system	操作系统
on top	在上面；领先；成功
boot loader	(计) 引导程序
device driver	(计) 设备驱动程序
malicious software	恶意软件
practical joke	恶作剧

Task Ⅲ-2

1. Match the following English phrases in Column A with their Chinese equivalents in Column B.

A	B
(1) boot loader	a. 应用软件
(2) assembly language	b. 系统软件
(3) system software	c. 汇编语言
(4) application software	d. 引导程序
(5) binary value	e. 恶意软件
(6) malicious software	f. 设备驱动程序

(7) practical joke　　g. 二进制数值
(8) each other　　h. 彼此
(9) device driver　　i. 操作系统
(10) operating system　　j. 恶作剧

2. Decide whether the following statements are T(true) or F (false) according to the passage.

(1) Computer software can be divided into application software and malicious software according to the passage. (　　)

(2) Application software is software that uses the computer system to perform special functions in the basic operation of the computer itself. (　　)

(3) There are many different types of application software, because the range of tasks that can be performed with a modern computer is so large. (　　)

(4) System software includes operating systems and device drivers. (　　)

(5) Operating systems are essential collections of software that manage resources and provides common services for other software. (　　)

(6) Supervisory programs, boot loaders and window systems are core parts of operating systems. (　　)

(7) In practice, an operating system comes bundled with additional software so that a user can potentially do some work. (　　)

(8) Device drivers operate or control a particular type of device that is attached to a computer. (　　)

(9) Utilities are computer programs designed to assist users in the maintenance and antivirus of their computers. (　　)

(10) Malware is software that is developed to harm and disrupt computers. (　　)

3. Translate the following passage into Chinese.

Operating systems are essential collections of software that manage resources and provides common services for other software that runs "on top" of them. Supervisory programs, boot loaders, shells and window systems are core parts of operating systems. In practice, an operating system comes bundled with additional software (including application software) so that a user can potentially do some work with a computer that only has one operating system.

Read the following passage and choose the best answer to each question.

Nature or Domain of Execution

Desktop applications such as web browsers and Microsoft Office, as well as smartphone and tablet applications (called "apps"). There is a push in some parts of the software industry to merge desktop applications with mobile apps, to some extent. Windows 8, and later Ubuntu Touch, tried to allow the same style of application user interface to be used on desktops, laptops and mobiles.

JavaScript scripts JavaScript scripts are pieces of software traditionally embedded in web pages that run directly inside the web browser when a web page is loaded without the need for a web browser plugin. Software written in other programming languages can also be run within the web browser if the software is either translated into JavaScript, or if a web browser plugin that supports that language is installed; the most common example of the latter is ActionScript scripts, which are supported by the Adobe Flash plugin.

Server software Server software includes web applications, which usually run on the web server and output dynamically generated web pages to web browsers, using e.g. PHP, Java, ASP. NET, or even JavaScript that runs on the server. In modern times these commonly include some JavaScript to be run in the web browser as well, in which case they typically run partly on the server, partly in the web browser.

Plugins and extensions Plugins and extensions are software that extends or modifies the functionality of another piece of software, and require that software be used in order to function.

Embedded software Embedded software resides as firmware within embedded systems, devices dedicated to a single use or a few uses such as cars and televisions (although some embedded devices such as wireless chipsets can "themselves" be part of an ordinary, non-embedded computer system such as a PC or smartphone). In the embedded system context there is sometimes no clear distinction between the system software and the application software. However, some embedded systems run embedded operating systems, and these systems do retain the distinction between system software and application software (although typically there will only be one, fixed, application which is always run).

Microcode Microcode is a special, relatively obscure type of embedded software which tells the processor itself how to execute machine code, so it is actually a lower level than machine code. It is typically proprietary to the processor manufacturer, and any necessary correctional microcode software updates are supplied by them to users (which is much cheaper than shipping replacement processor hardware). Thus an ordinary programmer would not expect to ever have to deal with it.

1. What are JavaScript scripts?
   A. They are pieces of software traditionally embedded in web pages that run directly inside the web browser when a web page is loaded with the need for a web browser plugin.
   B. They are pieces of software traditionally embedded in web pages that run indirectly inside the web browser when a web page is loaded with the need for a web browser plugin.
   C. They are pieces of software traditionally embedded in web pages that run directly inside the web browser when a web page is loaded without the need for a web browser plugin.

D. Software written in other programming languages can also be run within the web browser if the software is neither translated into JavaScript, nor if a web browser plugin that supports that language is installed.

2. Web applications usually run on the web server and output dynamically generated web pages to web browsers, using e. g. ________.

A. PHP that runs on the server.

B. JavaScript that runs on the server.

C. ASP. NET that runs on the server.

D. All of the above.

3. What are plugins and extensions according to the passage?

A. They are software that extends or modifies the functionality of another piece of software.

B. They are hardware that extends or modifies the functionality of another piece of software.

C. They are software that extends or modifies the functionality of another piece of hardware.

D. They are hardware that extends or modifies the functionality of another piece of hardware.

4. How does embedded software reside?

A. It resides as non-firmware within embedded systems, devices dedicated to only a single use such as cars.

B. It resides as firmware within non-embedded systems, devices dedicated to a single use or a few uses such as cars and televisions.

C. It resides as firmware within embedded systems, devices dedicated to a single use or a few uses such as cars and televisions.

D. It resides as firmware within embedded systems, devices dedicated to a single use or a lot of uses such as cars and televisions.

5. What is Microcode according to the passage?

A. It is a special, relatively obscure type of embedded software.

B. It tells the processor itself how to execute machine code, so it is a higher level than machine code.

C. It is typically proprietary to the processor users.

D. Any microcode software updates are supplied by users.

Section Four Writing

简历(Resume)

简历是个人情况介绍，一般附加在求职信后，目的是得到面试机会，因此简历要简洁具体，真实准确，侧重于工作经验和学历。

> Name：________
> Address：________
> Date of Birth：________
> Sex：Male / Female
> Marital Status：Single / Married / Divorced
> Job Objective：Seek a job as ________（工作种类）
> Education：________________________（倒序）
> Working Experience：________________________________（倒序）
> Foreign Languages：________________________________（语种及水平）

◆常用词语◆

① 名字（name ）

name in full(姓名)；surname/family name(姓)；first name/given name(名字)；middle name(中间名)

② 地址（address）

present/current address（现住址）；permanent address(永久联系地址)

③ 性别（sex/gender）

male（男）；female（女）

④ 婚姻状况(marital status)

single（单身）；unmarried（未婚）；married（已婚）；divorced（离婚）

⑤ job objective（求职意向）

a secretary to the General Manager(总经理秘书)；tourist guide(导游)；a position in marketing department(市场部的职位)；an assistant manager in a joint venture enterprise(合资企业的经理助理)

⑥ 受教育情况(education)

major(主修)；minor(辅修)；scores(分数)；grade(年级)；curriculum included(课程包括)；special training(特殊训练)；social practice(社会实践)；refresher course(进修课程)；extracurricular activities(课外活动)；scholarship(奖学金)；excellent leader(优秀学生干部)；"Three Goods" student("三好"学生)；B. A.（Bachelor of Arts）(文学学士)；B. S.（Bachelor of Science）（理学学士）；graduate student（研究生）；doctor（博士）；post doctorate(博士后)；M. B. A.（Master of Business Administration）(工商管理硕士)

⑦ foreign languages(外语水平)

fluent English speaking/interpreting/writing（能流利说/译/写英语）；English proficiency：CET Band IV（英语水平：通过大学英语四级考试）；fluent in speaking English and intermediate in reading Japanese(英语口语流利、日语阅读中等水平)

Task Ⅳ

1. You are required to complete the resume according to the information given.

请以中国学生张锦秋(女)的身份填写下列表格，具体信息如下：

出生年月：1982 年 10 月 16 日

联系地址：燕京市虹桥路 496 号

电话：000-67289879

个人详细情况：本人目前学历，学习英语的目的和经历，我的英语强项和弱项，希望在哪方面加强，等等。

Test Preparation Course Application Form

Family Name：________ First Name (s)：________

Date of Birth：________ Nationality：________

Sex：________ Telephone Number：________

Address：____________________

Please write about your current education, your strengths and weaknesses in English and your experience in English learning. (about 50 words)

____

____

____

2. You are required to complete the resume according to the information given.

姓名：李爱华，男，1980 年 5 月 16 日出生，未婚，家住北京市复兴路 61 号。2001 年以优异成绩毕业于华光技术学院计算机系。大学 3 年期间一直学习英语，有很好的阅读能力，具有用英语交流的能力。喜欢游泳和上网。欲求计算机程序员工作。

Resume

Name：________________________________

Address：________________________________

Date of birth：________________________________

Sex：male

Marital Status：single

Job objective：Seek a job as ________________________

Education：I graduated ________________________

Foreign languages：I studied English ________________________

Hobbies：________________________________

Words for reference

上网 surf the Internet

Unit Three Computer Network

Unit 3

Section One Speaking

Task Ⅰ

1. Look at the following graphic and speak out each part of Internet protocol suite.

Application

Transport

Network

Link(MAC)

Physical

Internet Protocol Suite ________ Physical ________

Link ________ Network ________

Transport ________ Application ________

2. List the Chinese names of the marks in the following network topology.

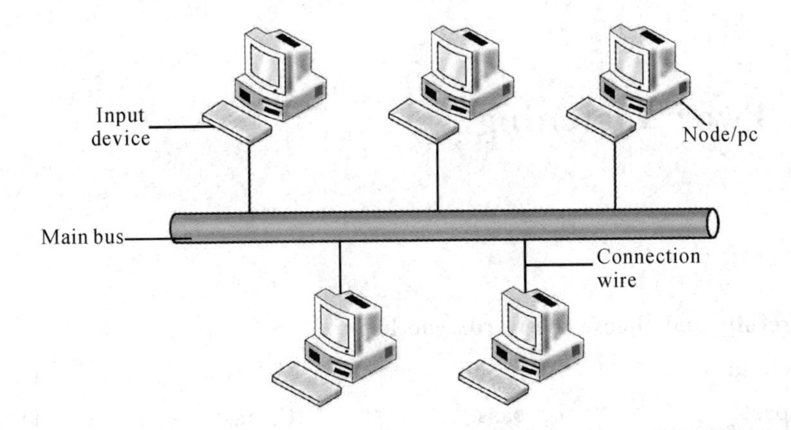

3. Suppose you are a network administrator. Give a presentation to each picture in English.

Picture 1. Introduce the common network devices to your customers.

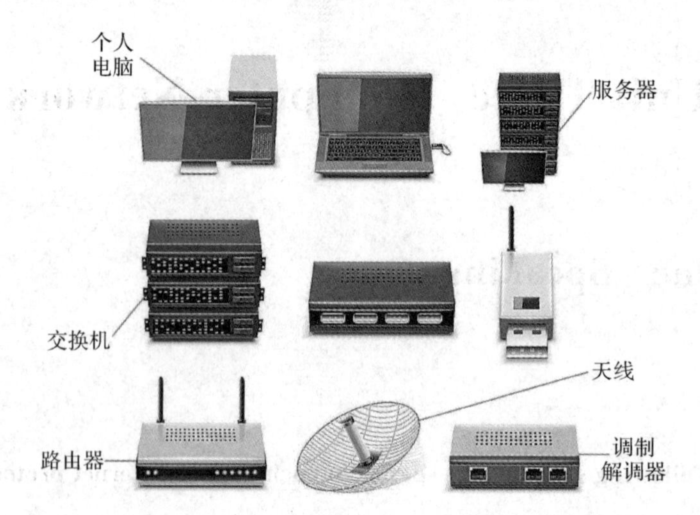

Picture 2. Tell your friends what you usually do in your daily work.

Section Two Listening

Task II

1. Listen carefully and choose the words you hear.

(1) A. cloud B. round C. loud D. ground

(2) A. past B. pass C. fast D. last

(3) A. father B. further C. leather D. weather

(4) A. shop B. shot C. shoot D. soft

(5) A. white	B. right	C. red	D. light
(6) A. book	B. boot	C. foot	D. fork
(7) A. shoes	B. whose	C. nose	D. mouse
(8) A. message	B. massage	C. methods	D. markets
(9) A. show	B. snow	C. whole	D. shoe
(10) A. sink	B. think	C. sick	D. thick

2. Listen to the short passage and choose the proper words to fill in the blanks.

GSM (Global System for Mobile) Communications is the most popular __1__ (qualification, standard, popularity) for mobile phones in the world. Its promoter, the GSM Association, estimates that __2__ (85%, 90%, 80%) of the global mobile market uses the standard. GSM is used by over 3 __3__ (billion, million) people across more than 212 countries and territories. Its ubiquity makes international roaming very common between mobile phone operators, enabling subscribers to use their phones in many parts of the world. GSM differs from its predecessors in that both signaling and speech channels are __4__ (digital, digitized, dignified), and thus is considered a second generation (2G) mobile phone system. This has also meant that __5__ (data, digital, stimulus) communication was easy to build into the system.

3. Listen to the dialogues and choose the best responses to what you hear.

(1) A. I'm flying back tomorrow.
B. See you next time.
C. It's my pleasure.
D. It's Tuesday. ________

(2) A. Speaking.
B. I'd like to put up a telephone service, please.
C. Is there Susan speaking?
D. You must have the wrong number. ________

(3) A. It's delicious.
B. It's lucky.
C. It's sure.
D. It's fine. ________

(4) A. Yes, I think so.
B. Do you really think so?
C. No, it's not very beautiful.
D. Oh, thank you. ________

(5) A. Maybe.
B. The same to you.
C. Where are you going?
D. OK. I'll visit my parents. ________

(6) A. Sorry, I'm a stranger here.

B. Well, I'll never know what you say.

C. Sorry, I won't give you the way.

D. Well, I'll tell you next time.

(7) A. Nice to see you.

B. How about you?

C. Not too bad.

D. How do you feel it?

(8) A. You are so polite.

B. We are so close friends.

C. Don't say so.

D. You are so welcome.

4. Listen to the dialogues and choose the best answer to each question you hear.

(1) A. Strict. B. Friendly.

C. Kind. D. Patient.

(2) A. A terrible traffic accident. B. A rush hour.

C. A terrible air crash. D. An airport.

(3) A. The pronunciation of the word. B. The spelling of the word.

C. The meaning of the word. D. The explanation of the word.

(4) A. Some coffee. B. Some juice.

C. Some cakes. D. Some oranges.

(5) A. Traveling. B. Reading stories.

C. Writing stories. D. Writing magazines.

5. Listen to the dialogue and answer the following questions by filling in the blanks.

(1) Who is Alice White?

She is a ______________.

(2) How old is she now?

She is ______________.

(3) When did she become famous?

She become famous ______________.

(4) Why has she given up swimming?

Because ______________.

(5) What time did she get up to go to?

She got up ______________.

6. Listen to the passage and fill in the blanks with words you hear. Some new words given below will be of some help to you.

communicate [kə'mjuːnikeɪt] vt. 传达，传送

cuddle ['kʌdl] v. 拥抱；搂抱

choppy ['tʃɔpɪ] adj. 不连贯的

overwhelm [ˌəuvə'welm] v. 压倒；击败；征服

console　[kən'səul]　v. 安慰；抚慰；慰藉
stimuli (单词原形：stimulus)　['stɪmulɪ]　n. 促进因素；激励因素
overload　[əuvə'ləud]　v. 使超载；使负荷过重
low-pitched　['ləupɪtʃt]　adj. (声音)低沉的
shout out　大叫，大声喊

How Babies Communicate?

Babies are born with the ___1___ to cry, which is how they ___2___ for a while. Your baby's cries generally tell you that something is ___3___: an empty belly, a wet bottom, cold feet, being tired, or a need to be held and cuddled, etc.

Soon you'll be able to ___4___ which need your baby is expressing and respond accordingly. In fact, sometimes what a baby needs can be identified by the type of cry-for example, the "I'm hungry" cry may be short and low-pitched, while "I'm upset" may ___5___ choppy.

Your baby may also cry when overwhelmed by all of the ___6___ and ___7___ of the world, or for no apparent reason at all. Don't be too ___8___ when your baby cries and you aren't able to console him or her ___9___: crying is one way babies shout out stimuli when they're ___10___.

Section Three　Reading

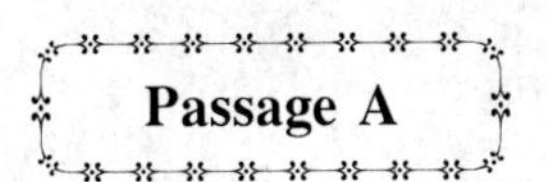

Passage A

Computer Network

A computer network or data network is a digital telecommunications network which allows nodes to share resources. In computer networks, networked computing devices exchange data with each other using a data link. The connections between nodes are established using either cable media or wireless media.

Network nodes

A computer or other device connected to a network, which has a unique address and is capable of sending or receiving data are called network nodes. Nodes can include hosts such as personal computers, phones, servers as well as networking hardware. Two such devices can be said to be networked together when one device is able to exchange information with the other device, whether or not they have a direct connection to each other.

Network Protocols

In order to get traffic flowing, you need protocols ruling how data is sent, received and interpreted by the networked devices. The protocols are stacked in several layers and together they make up the Internet Protocol Suite. On the lowest layer (the link layer) there is a protocol dealing with digital electronics. In a LAN or Local Area Network, this is the Ethernet Protocol. On top of this comes the Internet layer. Here we have the Internet Protocol (IP). From this point on a concept, referred to as IP addressing becomes available. Next in the Transport Layer, we find the Transmission Control Protocol or TCP for short, and User Datagram Protocol (UDP). Again, there may be other protocols in this layer. One more layer up is the Application Layer, which has protocols we know from everyday use, such as HTTP, SMTP and also P2P (peer to peer) voice communication.

Computer networks support an enormous number of applications and services such as access to the World Wide Web, digital video, digital audio, shared use of application and storage servers, printers, and fax machines, and use of email and instant messaging applications as well as many others. Computer networks differ in the transmission medium used to carry their signals, communications protocols to organize network traffic, the network's size, topology and organizational intent. The best-known computer network is the Internet.

★ ***New Words***

network [ˈnetwɜːk] n. 网络；广播网；网状物
telecommunication [ˌtelɪkəmjuːnɪˈkeɪʃ(ə)n] n. 远程通信；无线电通讯
node [nəʊd] n. 节点
networked [ˈnetwəːkt] adj. 网路的；广播电视联播的
exchange [ɪksˈtʃeɪndʒ; eks-] n. 交换；交流；交易所；兑换
connection [kəˈnekʃn] n. 连接；关系；人脉；连接件
establish [ɪˈstæblɪʃ; e-] vt. 建立；创办；安置
cable [ˈkeɪb(ə)l] n. 缆绳；电缆；海底电报
wireless [ˈwʌɪəlɪs] adj. 无线的；无线电的
host [həʊst] n. [计] 主机
protocol [ˈprəʊtəkɒl] n. 协议；草案；礼仪
interpret [ɪnˈtɜːprɪt] vt. 说明；口译 vi. 解释；翻译
stack [stæk] n. 堆；堆叠 vt. 使堆叠；把……堆积起来 vi. 堆积，堆叠
layer [ˈleɪə] n. 层，层次 vt. 把……分层堆放；
available [əˈveɪləb(ə)l] adj. 可获得的；可购得的；可找到的；有空的
concept [ˈkɒnsept] n. 观念，概念
transmission [trænzˈmɪʃ(ə)n] n. 传递；传送；播送
datagram [ˈdetəˌgræm] n. 数据电报
enormous [ɪˈnɔːməs] adj. 庞大的，巨大的
differ [ˈdɪfə] vt. 使……相异；使……不同 vi. 相异；意见分歧
medium [ˈmiːdɪəm] adj. 中等的；半生熟的 n. 方法；媒体；媒介
instant [ˈɪnst(ə)nt] adj. 立即的；紧急的；紧迫的 n. 瞬间；立即；片刻

signal ['sɪgnl] n. 信号；动机；标志
organizational [ˌɔːgənaɪ'zeɪʃənl] adj. 组织的；编制的
best-known ['best'nəun] adj. 流传久远的，最有名的

★ ***Phrases and Expressions***

computer network	计算机网络
data network	数据网络
data link	数据(自动)传输器
be capable of	能够
as well as	也；和……一样；不但……而且
Network Protocols	网络协议
traffic flowing	交通流量
make up	组成；补足；化妆；编造
Internet Protocol Suite	互联网协议组
link layer	链接层
digital electronics	数字电子技术
Local Area Network	局域网
Ethernet Protocol	以太网协议
Internet Protocol	互联网协议
refer to	参考；涉及；指的是；适用于
IP addressing	网络地址
transport Layer	传输层
transmission Control Protocol	传输控制协议
Application Layer	应用层
everyday use	日常使用
HTTP	超文本传输协议
SMTP	简单邮件传输协议
P2P (peer to peer) voice communication	点对点(对等点)语音通信
an enormous number of	大量的
access to	有权使用；通向……的入口
World Wide Web	万维网
digital video	数字视频
digital audio	数字音频
storage server	存储服务器
instant messaging	即时通讯
differ in	不同在；在……方面存在不同
transmission medium	传送介质
communications protocol	通信协议
network traffic	网络流量

Task Ⅲ - 1

1. Fill in the blanks without referring to the passage.

A computer network or data network is a digital telecommunications network which allows ________(1) to share ________(2). In computer networks, networked computing devices exchange ________(3) with each other using a ________(4). The connections between nodes are established using either ________(5) or ________(6).

2. Answer the following questions according to the passage.

(1) What is computer network?

____

(2) How to define network nodes? List several examples.

____

(3) Describe in your own words how the Internet Protocol Suite is organized.

____

(4) What is the top layer in the Internet Protocol Suite? Can you speak out several protocols we usually use in the layer?

____

(5) Which is the best-known computer network?

____

3. Complete each of the following statements according to the passage.

(1) A computer network or data network is a digital telecommunications network which allows ________ to share resources.

(2) In computer networks, networked computing devices exchange data with each other using a ________.

(3) A computer or other device connected to a network, which has a unique address and is capable of sending or receiving data are called ________.

(4) In order to get traffic flowing, you need ________ ruling how data is sent, received and interpreted by the networked devices.

(5) On the lowest layer (the link layer) there is a protocol dealing with ________.

(6) Computer networks differ in ________.

4. Translate the following sentences into English.

(1) 因特网是一个全球计算机网络系统。(network)

____

(2) 我们认为她完全能够照顾自己。(be capable of)

____

(3) 钱是用来买卖的一种媒介。(medium)

____

(4) 他的吉他弹得和你一样好。(as well as)

______

(5) 女警察构成警力的 13%。(make up)

______

(6) 唯一一点是不能用无线设备。(wireless)

______

(7) 到镇上唯一的通路是经过那座桥。(access to)

______

(8) 人们对失败持有不同的态度。(differ in)

______

5. Translate the following sentences into Chinese.

(1) A computer network or data network is a digital telecommunications network which allows nodes to share resources.

______

(2) In computer networks, networked computing devices exchange data with each other using a data link.

______

(3) A computer or other device connected to a network, which has a unique address and is capable of sending or receiving data are called network nodes.

______

(4) Two such devices can be said to be networked together when one device is able to exchange information with the other device, whether or not they have a direct connection to each other.

______

______

(5) The protocols are stacked in several layers and together they make up the Internet Protocol Suite.

______

(6) Next in the Transport Layer, we find the Transmission Control Protocol or TCP for short, and User Datagram Protocol (UDP).

______

(7) One more layer up is the Application Layer, which has protocols we know from everyday use.

______

(8) Computer networks differ in the transmission and communications protocols.

______

(9) The best-known computer network is the Internet.

______

(10) Computer networks support an enormous number of applications and services such as access to the World Wide Web.

____

Network Topology

Network topology is the arrangement of the various elements of a communication network, including its nodes and connecting lines. There are two ways of defining network geometry: the physical topology and the logical (or signal) topology.

The physical topology of a network is the actual geometric layout of workstations. There are several common physical topologies, as described below and as shown in the illustration.

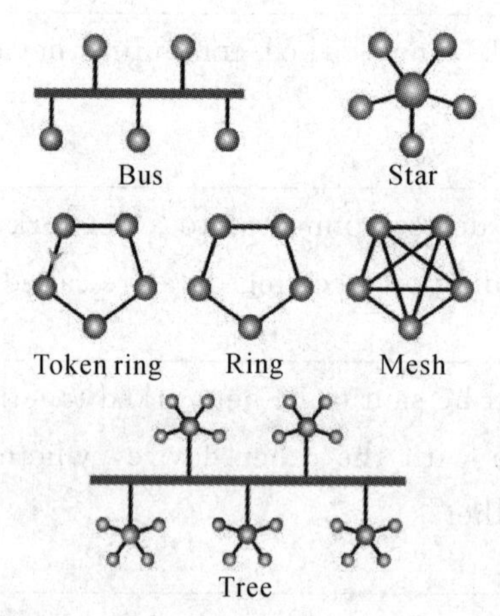

In the bus network topology, every workstation is connected to a main cable called the bus. Therefore, in effect, each workstation is directly connected to every other workstation in the network.

In the star network topology, there is a central computer or server to which all the workstations are directly connected. Every workstation is indirectly connected to every other through the central computer.

In the ring network topology, the workstations are connected in a closed loop configuration. Adjacent pairs of workstations are directly connected. Other pairs of workstations are indirectly connected, the data passing through one or more intermediate nodes.

If a Token Ring protocol is used in a star or ring topology, the signal travels in only one direction, carried by a so-called token from node to node.

The mesh network topology employs either of two schemes, called full mesh and partial mesh. In the full mesh topology, each workstation is connected directly to each of the others. In the partial mesh topology, some workstations are connected to all the others, and some are connected only to those other nodes with which they exchange the most data.

The tree network topology uses two or more star networks connected together. The central computers of the star networks are connected to a main bus. Thus, a tree network is a bus network of star networks.

Logical (or signal) topology refers to the nature of the paths the signals follow from node to node. In many instances, the logical topology is the same as the physical topology. But this is not always the case. For example, some networks are physically laid out in a star configuration, but they operate logically as bus or ring networks.

★ *New Words*

topology [tə'pɒlədʒɪ] n. 拓扑学
arrangement [ə'reɪn(d)ʒm(ə)nt] n. 布置；整理；准备
element ['elɪm(ə)nt] n. 元素；要素；原理
define [dɪ'faɪn] vt. 定义；使明确；规定
geometry [dʒɪ'ɒmɪtrɪ] n. 几何学，几何结构
logical ['lɒdʒɪk(ə)l] adj. 合逻辑的，合理的；逻辑学的
geometric [ˌdʒɪə'metrɪk] adj. 几何学的；[数] 几何学图形的
layout ['leɪaʊt] n. 布局；设计；安排；陈列
workstation ['wɜrksteɪʃn] n. 工作站
illustration [ɪlə'streɪʃ(ə)n] n. 说明；插图；例证；图解
bus [bʌs] n.（计算机）总线
central ['sentr(ə)l] adj. 中心的；主要的；中枢的 n. 电话总机
indirectly [ˌɪndɪ'rek(t)lɪ] adv. 间接地；不诚实地；迂回地
loop [lu:p] n. 环；圈 v. 使成环；以环连结
configuration [kənˌfɪgə'reɪʃ(ə)n] n. 配置；结构；外形
adjacent [ə'dʒeɪs(ə)nt] adj. 邻近的，毗连的
intermediate [ˌɪntə'mi:dɪət] adj. 中间的，中级的
direction [dɪ'rekʃn，daɪ-；] n. 方向；指导；趋势；用法说明
so-called [ˌsəʊ'kɔ: ld] adj. 所谓的；号称的
token ['təʊk(ə)n] n. 表征；代币；记号
mesh [meʃ] n. 网眼；网丝；圈套
scheme [ski:m] n. 计划；组合；体制；诡计
partial ['pɑ:ʃ(ə)l] adj. 局部的；偏爱的；不公平的
nature ['neɪtʃə] n. 自然；性质；本性；种类
path [pæθ] n. 道路；小路；轨道
physically ['fɪzɪkəllɪ] adv. 身体上，身体上地

★ *Phrases and Expressions*

connecting lines 连接线；外线

physical topology	物理拓扑
logical (or signal) topology	逻辑(或信号)拓扑
as shown in	如……所示
geometric layout	几何学的布置布线图
bus network topology	总线网络拓扑结构
be connected to	与……有联系；与……连接
in effect	实际上；生效
star network topology	星形网络
central computer	中央计算机
ring network topology	环形网络拓扑
pass through	穿过……；通过……
intermediate node	中间节点
mesh network topology	网状网络拓扑结构
full mesh	全网状
partial mesh	部分网状
tree network topology	树状网络拓扑结构
main bus	主总线
in many instances	在许多情况下
this is not always the case	情况并非一直如此
lay out	展示；安排；花钱；提议

Task Ⅲ - 2

1. Match the following English phrases in Column A with their Chinese equivalents in Column B.

A	B
(1) network topology	a. 中间节点
(2) connecting lines	b. 全网状网络拓扑
(3) physical topology	c. 逻辑拓扑
(4) logical topology	d. 环状网络拓扑
(5) star network topology	e. 物理拓扑
(6) central computer	f. 中央计算机
(7) ring network topology	g. 网络拓扑
(8) intermediate nodes	h. 主线
(9) mesh network topology	i. 星状网络拓扑
(10) main bus	j. 连接线

2. Decide whether the following statements are T(true) or F (false) according to the passage.

(1) Network topology is just the arrangement of the nodes. (　　)

(2) In the bus network topology, every workstation is connected to a main cable called

the bus. ( )

(3) In the bus network topology, each workstation is indirectly connected to every other workstation in the network. ( )

(4) In the star network topology, every workstation is indirectly connected to every other through the central computer. ( )

(5) In the ring network topology, all workstations are directly connected. ( )

(6) If a Token Ring protocol is used, the signal travels in only one direction, carried by a so-called token from node to node. ( )

(7) In the full mesh topology, each workstation is connected directly to each of the others. ( )

(8) In the partial mesh topology, not all workstations are connected to all the others. ( )

(9) All the computers of the star networks are connected to a main bus. ( )

(10) The logical topology is definitely different from the physical topology. ( )

3. Translate the following passage into Chinese.

Logical (or signal) topology refers to the nature of the paths the signals follow from node to node. In many instances, the logical topology is the same as the physical topology. But this is not always the case. For example, some networks are physically laid out in a star configuration, but they operate logically as bus or ring networks.

Read the following passage and choose the best answer to each question.

Wi-Fi

WiFi or Wi-Fi (/ˈwaɪfaɪ/) is a technology for wireless local area networking with devices based on the IEEE 802.11 standards.

Devices that can use Wi-Fi technology include personal computers, video-game consoles, phones and tablets, digital cameras, smart TVs, digital audio players and modern printers. Wi-Fi compatible devices can connect to the Internet via a WLAN and a wireless access point. Such an access point (or hotspot) has a range of about 20 meters (66 feet) indoors and a greater range outdoors. Hotspot coverage can be as small as a single room with walls that block radio waves, or as large as many square kilometers achieved by using multiple overlapping access points.

City-wide Wi-Fi

In the early 2000s, many cities around the world announced plans to construct citywide Wi-Fi networks. There are many successful examples; in 2004, Mysore became India's first Wi-Fi enabled city. A company called WiFiyNet has set up hotspots in Mysore, covering the complete city and a few nearby villages.

In 2005, St. Cloud, Florida and Sunnyvale, California, became the first cities in the United States to offer citywide free Wi-Fi (from MetroFi). Minneapolis has generated $1.2 million in profit annually for its provider.

In May 2010, London, UK, Mayor Boris Johnson pledged to have London-wide Wi-Fi by 2012. Several boroughs including Westminster and Islington already had extensive outdoor Wi-Fi coverage at that point.

Officials in South Korea's capital are moving to provide free Internet access at more than 10,000 locations around the city, including outdoor public spaces, major streets and densely populated residential areas. Seoul will grant leases to KT, LG Telecom and SK Telecom. The companies will invest $44 million in the project, which was to be completed in 2015.

Campus-wide Wi-Fi

Many traditional university campuses in the developed world provide at least partial Wi-Fi coverage. Carnegie Mellon University built the first campus-wide wireless Internet network, called Wireless Andrew, at its Pittsburgh campus in 1993 before Wi-Fi branding originated. By February 1997 the CMU Wi-Fi zone was fully operational. Many universities collaborate in providing Wi-Fi access to students and staff through the Eduroam international authentication infrastructure.

1. What is Wi-Fi?

   A. It is a physical device to catch the signal of internet.

   B. It is a technology for wireless local area networking.

   C. It is a central computer which can send signals to others.

   D. It is WLAN.

2. Which of the following devices can use Wi-Fi technology?

   A. Personal computers. B. Video-game consoles.

   C. Digital cameras. D. All of the above.

3. Which of the following cities is the earliest to construct citywide Wi-Fi networks?

   A. Mysore. B. London. C. Florida. D. California.

4. According to the passage, which of the following statements is true?

   A. Officials in Seoul intended to complete the whole city's Wi-Fi network in 2015.

   B. The officials in South Korea will charge people for using Wi-Fi.

   C. Officials in South Korea's capital are not moving to provide free Internet in densely populated residential areas.

   D. South Korea government has their own telecommunication company.

5. Carnegie Mellon University built the first campus-wide wireless Internet network in ______.

   A. 1997. B. 2012. C. 1993. D. 2015.

Section Four Writing

合同/协议(Contract / Agreement)

合同或协议是指为明确双方或多方在某一事务中的责任、义务和权利，在商务活动开始前，经相互讨论协商，双方或所涉及各方所签署的具有法律效应的书面正式文件。一般包括以下要素：

(1) 标题：要标明合同/协议性质。标题下是合同或协议参考号。

(2) 正文：包括开头、具体条款和结尾。开头写清合同或协议双方或各方全名和地址。一般在名称和地址后注上“以下简称甲/乙方、买/卖方或雇佣方/被雇佣方”，接着简要说明签订合同/协议的目的。具体条款要逐行排列，表明1、2、3等。结尾写明签署的时间、地点、份数及有效文字等。

(3) 签字：有关责任人签字或单位盖章。

CONTRACT

Date(日期)： Contract No.(合同号)：

The Buyers(买方)：

Address/TEL/FAX/E-mail(地址/电话/传真)：

The Sellers(卖方)：

Address/TEL/FAX/E-mail(地址/电话/传真)：

This contract is made by and between the Buyers and the Sellers; whereby the Buyers agree to buy and the Sellers agree to sell the under-mentioned goods subject to the terms and conditions as stipulated hereinafter(兹经买卖双方同意按照以下条款由买方购进，卖方售出以下商品)：

(1) Name of Commodity(商品名称)：

(2) Quantity(数量)：

(3) Unit price(单价)：

(4) Total Value(总值)：

(5) Packing(包装)：

(6) Country of Origin(生产国别)：

(7) Terms of Payment(支付条款)：

(8) insurance(保险)：

(9) Time of Shipment(装运期限)：

(10) Port of Lading(起运港)：

(11) Port of Destination(目的港)：

(12) Claims(索赔)：

Within 45 days after the arrival of the goods at the destination, should the quality, specifications or quantity be found not in conformity with the stipulations of the contract except those claims for which the insurance company or the owners of the vessel are liable, the Buyers shall, have the right on the strength of the inspection certificate issued by the C. C. I. C and the relative documents to claim for compensation to the Sellers.(在货到目的口岸 45 天内如发现货物品质，规格和数量与合同不附，除属保险公司或船方责任外，买方有权凭中国商检出具的检验证书或有关文件向卖方索赔换货或赔款。)

(13) Force Majeure(不可抗力)

The sellers shall not be responsible for the delay in shipment or non-delivery of the goods due to Force Majeure, which might occur during the process of manufacturing or in the course of loading or transit. The sellers shall advise the Buyers immediately of the occurrence mentioned above the within fourteen days there after. The Sellers shall send by airmail to the Buyers for their acceptance a certificate of the accident. Under such circumstances the Sellers, however, are still under the obligation to take all necessary measures to hasten the delivery of the goods.(由于人力不可抗力的缘由发生在制造、装载或运输的过程中导致卖方延期交货或不能交货者，卖方可免除责任，在不可抗力发生后，卖方须立即电告买方及在 14 天内以空邮方式向买方提供事故发生的证明文件，在上述情况下，卖方仍须负责采取措施尽快发货。)

(14) Arbitration(仲裁)

All disputes in connection with the execution of this Contract shall be settled friendly through negotiation. In case no settlement can be reached, the case then may be submitted for arbitration to the Arbitration Commission of the China Council for the Promotion of International Trade in accordance with the Provisional Rules of Procedure promulgated by the said Arbitration Commission. The Arbitration committee shall be final and binding upon both parties, and the Arbitration fee shall be borne by the losing parties.(凡有关执行合同所发生的一切争议应通过友好协商解决，如协商不能解决，则将分歧提交中国国际贸易促进委员会按有关仲裁程序进行仲裁，仲裁将是终局的，双方均受其约束，仲裁费用由败诉方承担。)

The Buyers(买方)：　　　　The Sellers(卖方)：

◆常用词语◆

① Party A and Party B agree to the following terms after friendly negotiations based on the principles of mutual interest and equal rights.(甲、乙双方经友好协商，在互惠互利、平等公正基础上，达成以下协议。)

② If any terms need to be altered, the two parties shall sign a new supplemental agreement, which shall have the same force as this agreement.(其他未尽事宜，双方协商解决并另定补充协议。)

③ This lease is written in both Chinese and English, and both of which are authentic.

(本合同以中、英两种文字写成，具同等法律效力。)

④ The two parties shall settle disputes through friendly negotiations.（如有争议，双方应协商解决。）

⑤ This agreement has two original copies，one for each party.（本合同文本一式两份，甲、乙双方各执一份。）

⑥ The two parties shall strictly follow the items in this agreement.（双方应严格遵守本合同所有条款。）

⑦ This agreement is effective from the date when both Party A and Party B have signed it.（本合同自双方签署之日起生效。）

Task Ⅳ

You are required to write a Telephone Message according to the following instructions given in Chinese.

名称:销售合同

正文:本合同由加拿大CBD公司和中国DL公司通过友好协商签订，条款如下。

(1) 货物名称：玉米

(2) 数量：2万吨

(3) 单价：4美元/公斤

(4) 总值：8000万美元

(5) 包装：货物应具有防潮、防震并适合远洋运输的包装

(6) 本合同以中、英文两种文字写成，具有同等法律效力

(7) 原产地：美国

(8) 装运期限：4月10日前

(9) 装运口岸：美国纽约

(10) 目的口岸：中国大连

(11) 保险：由卖方按发票金额110%投保一切风险

(12) 付款方式：信用证付款

Unit 4

Unit Four Multimedia

Section One Speaking

Task Ⅰ

1. Look at the following pictures and speak out each name of them in Chinese.

A. projector ________ B. ipad C. camera ________

D. chassis ________ E. printer ________

2. In modern society, multimedia information can be found everywhere. In movies, we use it to make special effects; in business, we use it to sell ideas or products. Where else has it been used around you? Can you give some examples?

3. Suppose you are a salesman in Fox Technology Corporation which specializes in selling multimedia classroom devices. Give a presentation to each picture in English.

Picture 1

Picture 2

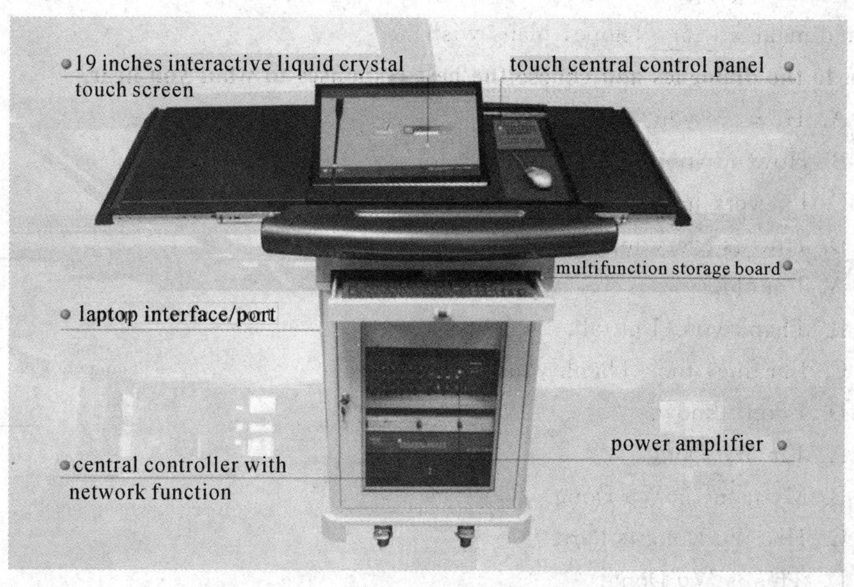

Section Two Listening

Task Ⅱ

1. Listen carefully and choose the words you hear.

(1) A. tea B. great C. meat D. heat
(2) A. shout B. house C. mouse D. low
(3) A. watched B. stopped C. cooked D. played
(4) A. happy B. easy C. why D. city
(5) A. horse B. orange C. short D. or
(6) A. girl B. worker C. skirt D. shirt
(7) A. wait B. play C. day D. Sunday
(8) A. butter B. teacher C. driver D. her
(9) A. yellow B. know C. blow D. how
(10) A. ear B. heard C. tear D. appear

2. Listen to the short passage and choose the proper words to fill in the blanks.

For many people, a __1__ (birthday, New Year, meeting) is one of the most important days of the year. And it is time for celebrations with __2__ (families, friends, families and friends). On this day, they have made many delicious food, such as ice-cream, cake and __3__ (meat, sandwiches, fish). They can also receive many gifts. __4__ (Opening, closing, putting) the gifts is the nicest part of the party. They are singing, dancing and having a good time. Before they eat the cake, the birthday child will blow out candles and make a __5__ (hope, plan, wish).

3. Listen to the dialogues and choose the best responses to what you hear.

(1) A. How are you?
B. How do you do?
C. I'm very happy.
D. How is everything going? ______

(2) A. I'm old.
B. Thank you. I'm tall.
C. I'm fine, too. Thank you.
D. I don't know. ______

(3) A. I'm Wu Dong.
B. My name is Wu Dong.
C. Hi, Wu Dong is here.
D. This is Wu Dong. ______

(4) A. There are some books in it.
B. They are books.
C. It has books in it.
D. Books are in it. ______

(5) A. For 2 hours.
B. One time a week.
C. Twice a week.
D. Two times a week. ______

(6) A. Glad to see you.
B. With pleasure.
C. Yes, you are right.
D. I want to help you. ______

(7) A. Nothing is wrong.
B. No reason.
C. It's nothing.
D. No problem. ______

(8) A. Yes, I like.
B. Yes, I do.
C. Yes, I'd love to.
D. Yes, love. ______

4. Listen to the dialogues and choose the best answer to each question you hear.

(1) A. Buy a bookshelf. B. Sell a bookshelf.
C. Repair a bookshelf. D. Rearrange the furniture.

(2) A. By plane. B. By bus.

C. By ship. D. On foot.

(3) A. At the drugstore. B. At the department store.

C. At the airport. D. At the grocery store.

(4) A. Baby sitter. B. Teacher.

C. Nurse. D. Doctor.

(5) A. At 7：00. B. At 6：30.

C. At 7：30. D. At 6：00.

5. Listen to the dialogue and answer the following questions by filling in the blanks.

(1) Where does the conversation take place?

It takes place in the ____________.

(2) What would Mr. Wang like to drink?

He likes to drink ____________.

(3) What's the relationship between the two speakers?

They are ____________.

(4) Mr. Wang also likes to eat something delicious. What is it?

It is a ____________.

(5) What would Mr. Wang like to go with his steak?

He will have a green ____________.

6. Listen to the passage and fill in the blanks with words you hear. Some new words given below will be of some help to you.

noodle	[ˈnuːdl]	n.	面条
freshly	[ˈfreʃli]	adv.	气味清新地
hamburger	[ˈhæmbəːgə]	n.	汉堡
habit	[ˈhæbit]	n.	习惯
salad	[ˈsæləd]	n.	沙拉
fork	[fɔːk]	n.	叉
chopstick	[ˈtʃɔpstik]	n.	筷子
rude	[ruːd]	adj.	粗鲁的；不礼貌的
during	[ˈdjuəriŋ]	prep.	(表示时间)在……期间
have dinner			进餐

Different eating habits

Different countries have different eating __1__. And I'd like to say something about it.

Firstly, Chinese people eat a lot of __2__ rice, noodles, __3__, meat, and __4__. And the food is often cooked in many ways. Western people prefer bread, milk and __5__. They also eat a lot of fast food like hamburgers and __6__. Secondly, Westerners are used to eating meals with a __7__ and a __8__, and Chinese use __9__. Thirdly, Westerners think it's __10__ to make noise during the meal, but it is common for Chinese to talk with friends when having dinner.

Section Three Reading

Passage A

Multimedia

What is Multimedia

Multimedia is content that uses a combination of different content forms such as text, audio, images, animations, video and interactive content. A perfect example of multimedia is an English learning platform that has text regarding the learning points, the audio files of the passages, and even a video of the contents being performed in real situations.

Multimedia applications

Multimedia finds its applications in various areas including, but not limited to, advertisements, art, education, entertainment, engineering, medicine, business and scientific research. Several examples are as follows:

Commercial uses

Much of the electronic old and new media used by commercial artists and graphic designers is multimedia. Exciting presentations are used to grab and keep attention in advertising. Business to business, and interoffice communications are often developed by creative services firms for advanced multimedia presentations beyond simple slide shows to sell ideas or liven up training. Commercial multimedia developers may be hired to design for governmental services and nonprofit services applications as well.

Entertainment

Multimedia is heavily used in the entertainment industry, especially to develop special effects in movies and animations. Multimedia games are also a popular pastime and are software programs available either as CD-ROMs or online. Some video games also use multimedia features. Multimedia applications that allow users to actively participate instead of just sitting by as passive recipients of information are called interactive multimedia. In the arts there are multimedia artists, whose minds are able to blend techniques using different media that in some way incorporates interaction with the viewer. One of the most relevant could be Peter Greenaway who is melding cinema with opera and all sorts of digital media.

Education

In education, multimedia is used to produce computer-based training courses (popularly called CBTs) and reference books like encyclopedia and almanacs. A CBT lets the user go through a series of presentations, text about a particular topic, and associated illustrations in various information formats. Edutainment is the combination of education with entertainment，especially multimedia entertainment.

The future of multimedia

In the science fiction “Neuromancer”, William Gibson describes a space, Cyberspace, controlled by a computer. Once his brain was linked with the computer, a man would undergo all experiences in the space. His various senses in the realistic world would be replaced with a series of new electronic stimuli. The Cyberspace is regarded as a goal of future virtual reality which integrates the sense of touch with video and audio media to immerse an individual into a virtual world. For example, in a virtual space, students can “dissect” a human body, “visit” ancient battlefields, or “talk” with Shakespeare.

★ *New Words*

text [tekst] n. [计] 文本
audio ['ɔ:dɪəʊ] adj. 声音的
animation [ænɪ'meɪʃ(ə)n] n. 活泼，生气；激励；卡通片
interactive [ɪntər'æktɪv] adj. 交互式的
platform ['plætfɔ:m] n. 平台；月台
perform [pə'fɔ: m] v. 执行；履行；表演；扮演
entertainment [ˌentə'teɪnmənt] n. 娱乐
commercial [kə'mɜ:ʃ(ə)l] adj. 商业的
presentation [prez(ə)n'teɪʃ(ə)n] n. 展示；描述
grab [græb] v. 攫取；霸占
advertising ['ædvətaɪzɪŋ] n. 广告；广告业
advanced [əd'vɑ:nst] adj. 先进的；高级的
governmental [ˌgʌvn'mentl] adj. 政府的；政治的
nonprofit [nɒn'prɒfɪt] adj. 非营利的
pastime ['pɑ:staɪm] n. 娱乐；消遣
actively ['æktivli] adv. 积极地；活跃地
participate [pɑ:'tɪsɪpeɪt] vi. 参与，参加；分享　vt. 分享；分担
passive ['pæsɪv] adj. 被动的，消极的
recipient [rɪ'sɪpɪənt] n. 收件人
incorporate [ɪn'kɔ:pəreɪt] vt. 包含，吸收；体现；把……合并
interaction [ɪntər'ækʃ(ə)n] n. 相互作用
relevant ['reləvənt] adj. 相关的；切题的；中肯的
meld [meld] v. 使……合并；使……混合　n. 结合；融合
reference ['ref(ə)r(ə)ns] n. 参考，参照
encyclopedia [ɪnˌsaɪklə'pi:diə] n. 百科全书
almanacs ['ɔ:lmənæk; 'ɒl-] n. 年鉴；历书
associated [ə'soʃɪetɪd] adj. 关联的；联合的
format ['fɔ: mæt] n. [计] 格式

realistic [rɪə'lɪstɪk] adj. 现实的
integrate ['ɪntɪgreɪt] vt. 使……完整；使……成整体
immerse [ɪ'mɜːs] vt. 沉浸；使陷入
dissect [daɪ'sekt; dɪ-] vt. 切细；仔细分析 vi. 进行解剖
battlefield ['bæt(ə)lfiːld] n. 战场

★ ***Phrases and Expressions***

a combination of	事物的结合；综合……
interactive content	互动式内容
but not limited to	但不限于
graphic designer	平面设计师
slide shows	放映幻灯片
liven up	使……有生气
special effects	特技效果
multimedia artist	多媒体艺术家
computer-based training courses	计算机辅助培训
go through	参加，经受，通过
a series of	一系列的
science fiction	科幻小说
linked with	与……有关，与……相连接
be replaced with	替换为，以……代替
be regarded as	被认为是，被当作是
virtual reality	虚拟现实
sense of touch	触觉

Task Ⅲ - 1

1. Fill in the blanks without referring to the passage.

Multimedia is content that uses a ________(1) of different content forms such as ________(2), ________(3), ________(4), animations, video and ________(5) content. A perfect example of multimedia is an English learning platform that has text regarding the learning points, the audio files of the passages, and even a video of the contents being ________(6) in real situations.

2. Answer the following questions according to the passage.

(1) What is multimedia?

____

(2) Could you please give us an example of multimedia?

____

(3) In which areas can multimedia be used? Can you give some examples?

____

(4) Is a simple slide show enough to sell ideas or liven up trainings?

____

(5) Do you think Virtue Reality can promote the development of education? Why or why not?

____

____

3. Complete each of the following statements according to the passage.

(1) Multimedia is content that uses a ________ of different content forms such as text, audio, images, animations, video and interactive content.

(2) Multimedia finds its ________ in various areas including, but not limited to, advertisements, art, education, entertainment, engineering, medicine.

(3) Business to business, and interoffice communications are often developed by creative services firms for ________ presentations beyond simple slide shows to sell ideas or liven up training.

(4) Multimedia is heavily used in the entertainment industry, especially to develop ________ in movies and animations.

(5) In education, multimedia is used to produce ________ training courses (popularly called CBTs) and reference books like encyclopedia and almanacs.

(6) The Cyberspace is regarded as a goal of future ________ which integrates the sense of touch with video and audio media to immerse an individual into a virtual world.

4. Translate the following sentences into English.

(1) 然后我会得出它们的一个组合。(a combination of)

____

(2) 但不仅仅是这个，还有更深的含义。(but not limited to)

____

(3) 这篇论文介绍了一种多媒体数据库引擎。(multimedia)

____

(4) 你要过境，就必须在海关办理手续。(go through)

____

(5) 这座桥将该岛和大陆连接起来。(linked with)

____

(6) 这个小镇提供各种各样的娱乐活动。(entertainment)

____

(7) 往日的学校已被那座高楼取代。(be replaced with)

____

(8) 参加这个会议的有非营利性的团体、国际健康专家和非洲领袖。(nonprofit)

____

5. Translate the following sentences into Chinese.

(1) Multimedia is content that uses a combination of different content forms such as

text, audio, images, animations, video and interactive content.

____

(2) A perfect example of multimedia is an English learning platform that has text regarding the learning points, the audio files of the passages, and even a video of the contents being performed in real situations.

____

(3) Multimedia finds its application in various areas including, but not limited to, advertisements, art, education and entertainment.

____

(4) Exciting presentations are used to grab and keep attention in advertising.

____

(5) Business to business, and interoffice communications are often developed by creative services firms.

____

(6) Multimedia is heavily used in the entertainment industry, especially to develop special effects in movies and animations.

____

(7) Multimedia applications that allow users to actively participate instead of just sitting by as passive recipients of information are called interactive multimedia.

____

(8) In education, multimedia is used to produce computer-based training courses (popularly called CBTs).

____

(9) Once his brain was linked with the computer, a man would undergo all experiences in the space.

____

(10) The Cyberspace is regarded as a goal of future virtual reality which integrates the sense of touch with video and audio media to immerse an individual into a virtual world.

____

____

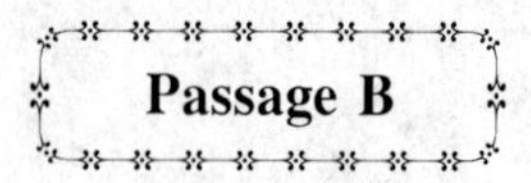

How to Create a Multimedia Presentation

A multimedia presentation differs from a normal presentation in that it contains some forms of animation or media. Typically a multimedia presentation contains at least one of the following elements:

Video or movie clip

Animation

Sound (this could be a voice-over, background music or sound clips)

Navigation structure

Choice of multimedia presentation technology

The first-and hardest-part is to choose the technology for your presentation. The choice comes down to two main contenders, Adobe Flash or Microsoft PowerPoint.

Adobe Flash

Flash allows you to create presentations where you can build in powerful animation. It also has very good video compression technology.

Perhaps the best part of Flash is that it also allows you to put presentations directly onto your web site.

The biggest problem though is that Flash is a difficult system to get to use. I have been on a training class and also have access to a couple of graphic designers for help and still find it difficult to put together a presentation in Flash.

Life has become a lot easier in the recent versions (Flash 8 and Flash CS3). With these versions there is a new feature called Flash Slide Presentation. Rather than the conventional time-line it allows you to build and add in slides a bit like the slide sorter in PowerPoint.

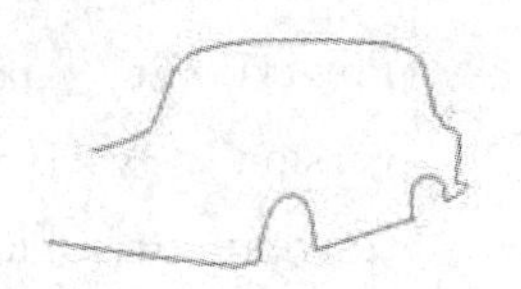

It is also very expensive. Checking on Amazon the latest version of Flash (CS3) will set you back £515 ($629.99). You will probably have to factor in around £400 for a decent Flash training class.

There is a low cost alternative to Flash called Swish. They make it easy to build a Flash presentation without the need for detailed programming knowledge. For a free trial of the software you can visit the Swish website at www.swishzone.com.

Microsoft PowerPoint

The easiest way to create a multimedia presentation is in Microsoft PowerPoint. You can add in video, a soundtrack and also an animation.

By far the biggest advantage of making multimedia presentations in PowerPoint is that it is easy for anyone to be able to edit the presentation.

An example of how you can introduce multimedia effects in a presentation is shown in the taxi sequence on the right. It starts with a line drawing being made on the screen and is then followed up with a taxi fading in. The sequence is part of one of our PowerPoint templates. You can see the effect in action by downloading the taxi animated template from the download page. Make sure that you view the sequence in PowerPoint show mode.

We have a number of tutorials on the Presentation Helper web site that make it easier to put together a multimedia presentation.

How to add video clips into a PowerPoint presentation

How to add music into PowerPoint

Simple menus

How to animate your slides

If you put together a number of these elements you will have a way to create a multimedia presentation in PowerPoint.

★ *New Words*

voice-over [ˈvɔisˌəuvə] n. 画外音；(电影或电视)旁白

navigation [nævɪˈgeɪʃ(ə)n] n. 航行；航海

contender [kənˈtendə] n. 竞争者；争夺者

PowerPoint [ˈpauəpɔint] n. 微软办公软件

version [ˈvɜːʃ(ə)n] n. 版本；译文；倒转术

conventional [kənˈvenʃ(ə)n(ə)l] adj. 符合习俗的，传统的；常见的；惯例的

time-line [ˈtaɪmˌlaɪn] n. 时间线；等时线

sorter [ˈsɔrtə] n. 从事分类的人；分类机

slide [slaɪd] n. 滑动；幻灯片；滑梯；雪崩 v. 滑动

decent [ˈdiːs(ə)nt] adj. 正派的；得体的；相当好的

programming [ˈprəʊgræmɪŋ] n. 设计，规划；[计] 程序编制

soundtrack [ˈsaʊn(d)træk] n. 声带；声道；声迹；电影配音

edit [ˈedɪt] vt. 编辑；校订 n. 编辑工作

sequence [ˈsiːkw(ə)ns] n. [数][计] 序列；顺序；续发事件 vt. 按顺序排好

template [ˈtempleɪt; -plɪt] n. 模板，样板

tutorial [tjuːˈtɔːriəl] n. 导师辅导(时间)，软件教程

animate [ˈænɪmeɪt] vt. 使有生气；使活泼；鼓舞；推动 adj. 有生命的

★ *Phrases and Expressions*

differ from	与……不同；区别于……
in that	因为；由于；既然
movie clip	影片剪辑
background music	背景音乐
come down to	归根结底，可归结为；实质上是

Adobe Flash	Flash 动画制作软件
build in	插入；嵌入
video compression	视频压缩
have access to	使用；接近；可以利用
add in	添加；把……包括在内
set you back	让你破费
factor in	将……纳入；把……计算在内
alternative to	可供选择 可供选择的替代
free trial	[贸易] 免费试用
line drawing	素描；线条画
follow up	跟踪；坚持完成；继续做某事
fade in	淡入；渐显
in action	在活动；在运转

Task Ⅲ-2

1. Match the following English phrases in Column A with their Chinese equivalents in Column B.

A	B
(1) multimedia presentation	a. 背景音乐
(2) movie clips	b. 低成本
(3) voice-over	c. 多媒体演示文稿
(4) background music	d. 下载区
(5) navigation structure	e. PowerPoint 模板
(6) video compression	f. 视频剪辑
(7) low cost	g. 影片剪辑
(8) PowerPoint templates	h. 视频压缩
(9) download page	i. 导航结构
(10) video clips	j. 画外音；旁白

2. Decide whether the following statements are T(true) or F (false) according to the passage.

(1) A multimedia presentation differs from a normal presentation in that it contains some form of animation or media. (　　)

(2) Typically a multimedia presentation must contain the following elements: video or movie clip, animation, sound and navigation structure. (　　)

(3) Adobe Flash and Microsoft PowerPoint are two main contenders for making multimedia presentation. (　　)

(4) Adobe Flash has very good video compression technology. (　　)

(5) The best part of PowerPoint is that it also allows you to put presentations directly onto your web site. (　　)

(6) The biggest problem though is that Flash is too expensive. (　　)

(7) Swish is cheaper than Flash. (　　)

(8) You can add in video, a soundtrack and also an animation when you create a multimedia presentation in Microsoft PowerPoint. (　　)

(9) The biggest advantage of making multimedia presentations in PowerPoint is that it is easy for anyone to be able to edit the presentation. (　　)

(10) You can view the animation in any mode of PowerPoint. (　　)

3. Translate the following passage into Chinese.

The biggest problem though is that Flash is a difficult system to get to use. I have been on a training class and also have access to a couple of graphic designers for help and still find it difficult to put together a presentation in Flash.

Life has become a lot easier in the recent versions (Flash 8 and Flash CS3). With these versions there is a new feature called Flash Slide Presentation. Rather than the conventional time-line it allows you to build and add in slides a bit like the slide sorter in PowerPoint.

Extra Reading

Read the following passage and choose the best answer to each question.

Adobe Authorware

Authorware was originally produced by Authorware Inc., founded in 1987 by Dr Michael Allen. Allen had contributed to the development of the PLATO computer-assisted instruction system during the 1970s that was developed jointly by the University of Illinois and Control Data Corporation.

Authorware became a rapid success in the marketplace, obtaining more than 80% of the market in about three years. Authorware Inc. merged with MacroMind/Paracomp in 1992 to form Macromedia. In December 2005, Adobe and Macromedia merged, under the Adobe Systems name.

Adobe Authorware (previously Macromedia Authorware) was an interpreted, flowchart-based, graphical programming language. Authorware is used for creating interactive programs that can integrate a range of multimedia content, particularly electronic educational technology (also called e-learning) applications. The flowchart model differentiates Authorware from other authoring tools, such as Adobe Flash and Adobe Director, which rely on a visual stage, time-line and script structure.

Authorware used a visual interface (可视化界面) with icons, representing essential components of the interactive learning experience. "Authors" placed icons along a "flowline" to create a sequence of events. Icons represented such components as Display—put something on the screen, Question—ask the learner for a response, Calc—perform a calculation, read data, and/or store data, and Animate — move something around on the screen. By simply placing the icons in sequence and adjusting their properties, authors could instantly see the structure of program they were creating and, most importantly, run it to see what learners would see. On-screen changes were easy to make, even while the program was running.

Authorware is particularly well suited to creating electronic educational technology (also called e-learning) content, as it includes highly customizable templates for CBT and WBT, including student assessment tools. Working with these templates, businesses and schools can rapidly assemble multimedia training materials without needing to hire a full-fledged programmer. Intuitively-named dialog boxes take care of input and output. The flow chart model makes the re-use of lesson elements extremely straightforward. Being both AICC-and SCORM-compliant, Authorware can be used to deliver content via any AICC or SCORM Learning Management System.

Moving beyond the templates, however, requires either the importing of interactive Flash or Director movies, or scripting, which can be done in Authorware's native scripting language or in JavaScript.

1. Authorware was originally produced by ________.
   A. University of Illinois.　　B. Authorware Inc.
   C. Control Data Corporation.　　D. MacroMind.
2. The name Adobe Systems was fixed in ________.
   A. 1970s　　B. 1987　　C. 1992　　D. 2005
3. The underlined word"integrate"(in Paragraph 3)probably means ________.
   A. 求积分　　B. 使……成整体
   C. 取消隔离　　D. 完全的
4. Which of the following icons have been referred to according to the passage?
   A. Display—put something on the screen.
   B. Question—ask the learner for a response.
   C. Calc—perform a calculation.
   D. All the above.
5. It can be inferred from the text that ________.
   A. Authorware can be widely used in education field
   B. working with the templates, businesses and schools need to hire a full-fledged programmer to assemble multimedia training materials rapidly
   C. the flow chart model are not necessary for the re-use of lesson element
   D. It is hard for Authorware to deliver content via any AICC-or SCORM-Learning Management System

Section Four Writing

证明信和致辞(Certificate& Speech)

1. 证明信(Certificate)

证明信是个人所在单位、学校或相关人员、部门出具的证明个人身份、学习经历、工作经历或其它一些事实真相的一种文书，具有一定的法律效力。证明信可分为组织证明信(介绍信)和个人证明信。前者又可分为普通书写证明信和印刷证明信。证明信通常采用一般信件的格式，称呼多用:"To whom it may concern"，有时也可省略，在很多情况下，会省去收信人的姓名和地址。

July 12, 2016

To Whom It May Concern,

This is to certify that ________________(被证明人的身份、成绩、能力等。). The personal materials will be provided by himself/herself.

Certified true and correct!

Yours faithfully,

________(签名)

________(职务)

________(单位)

◆常用语句◆

① Identification Certificate/Birth Certificate/Certificate of Physical Examination(身份证明、出生证明、体检证明)

② To whom it may concern (致有关负责人)

③ This is to certify/It is my pleasure to certify that Mrs. Wang Ling, one of my former students, is a teaching assistant in our university now.(兹证明王灵女士是我从前的学生，现在的我校助教。)

④ This is to certify that... completed four years of study at... university and graduated on...(兹证明……在……大学完成四年学业，并于……毕业。)

⑤ This is to certify that... has been employed in our company as.. for the past five years. 兹证明……过去5年在我公司担任……

⑥ Being physically and mentally healthy, he has faithfully attended to his duties and has proved himself to be industrious and thoroughly reliable.(他身体健康，工作努力，忠于职责，诚实可靠。)

⑦ Wang Qi got excellent scores in all the subjects. In addition，he was on good terms with his classmates.（王齐在校期间成绩优秀，与同学相处融洽。）

⑧ Certified true and correct!（证明真实无误!）

2. 致词(Speech)

致词是指在正式、隆重的场合，主持人或出席人员为某目的而发表的讲话，如开幕词、闭幕词、欢迎词、欢送词、告别词、祝酒词等。一般包括称呼、讲话内容和祝愿结束语三部分。

> Ladies and Gentlemen，
>
> Good morning to everyone，and thank you，for joining us at the ________. It gives me a great pleasure on behalf of our university to extend a warm welcome to ________.
>
> ____
>
> ____.
>
> Thank you all and wish you a good time.

◆致辞常用语句◆

① On behalf of sb，I have the honor to extend this warm welcome to sb.（我很荣幸地代表……向……表示热烈的欢迎。）

② It is my honour to declare the conference of... open.（我很荣幸地宣布……大会开幕了!）

③ It gives me a great pleasure on behalf of our university to extend a warm welcome to Professor Brown.（非常高兴能代表我校向布朗教授的到来表示热烈的欢迎。）

④ I would like to share with you brief information.（在此，我愿意和朋友们分享一些简要信息。）

⑤ Wish a complete success.（预祝取得圆满成功。）

⑥ I am deeply grateful for everything you've done for me since my arrival in China.（我对您在我到中国后所做得一切安排表示深深的谢意。）

⑦ Before leaving，it gives me a great pleasure to express thanks earnestly to you for your warm reception and hospitality extended to me generously during my visit.（临别之际，我非常高兴并诚挚地向诸位表示感谢，感谢诸位在访问期间给予的热情而慷慨的款待。）

⑧ I wish to propose a toast：To the health of sb. /To the new stage of sth.（我提议为某人的健康/为某事开始新篇章而干杯!）

Task Ⅳ

1. You are required to write a certificate according to the following instructions given in Chinese.

在读证明

谢韩同学，男，1999年9月25日生，2015年9月1日作为高中学生来沈阳育才中学就读，现高二(十)班在读生。

特此证明！

育才中学教学管理办公室

2016.8.10

Words for reference:

高中学生：senior high school student

育才中学：Yucai High School

教学管理办公室：Office of Educational Administration

2. You are required to write a farewell speech according to the information given in Chinese.

假设你是政府工作人员，去德国考察时，参观了一些工厂、学校和文化团体，与他们的政府官员、科学家、工人、教师、学生进行了交谈，并与他们成了朋友。临别前你参加了他们为你举行的欢送晚会。请你就上述内容写一篇临别致辞。

Unit Five E-commerce

Section One Speaking

Task Ⅰ

1. Look at the following diagram and speak out the common types of E-commerce in Chinese.

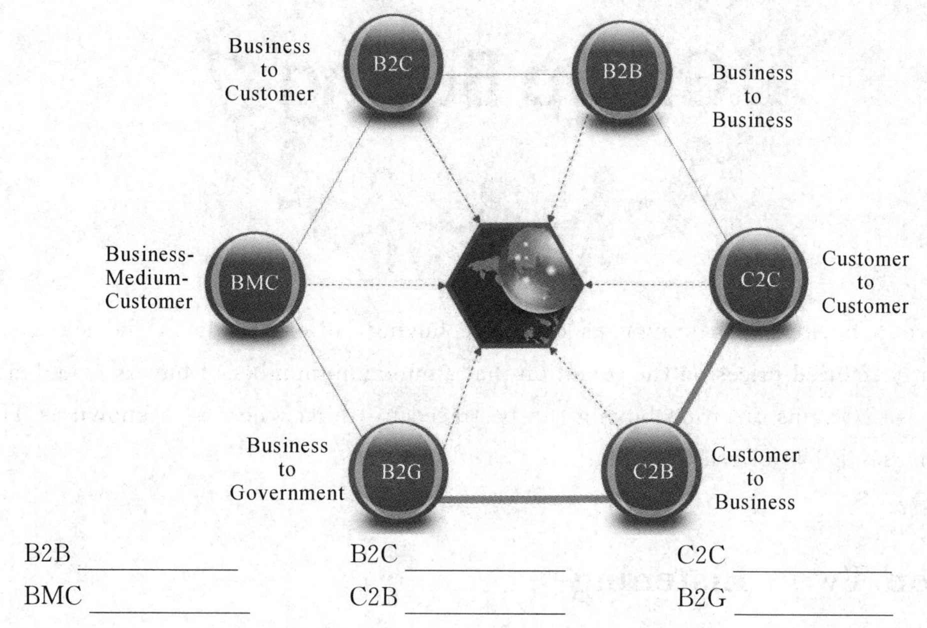

B2B ____________ B2C ____________ C2C ____________

BMC ____________ C2B ____________ B2G ____________

2. The flow chart below describes the basic process of E-commerce in detail. Look at it closely and communicate your opinions with your partner.

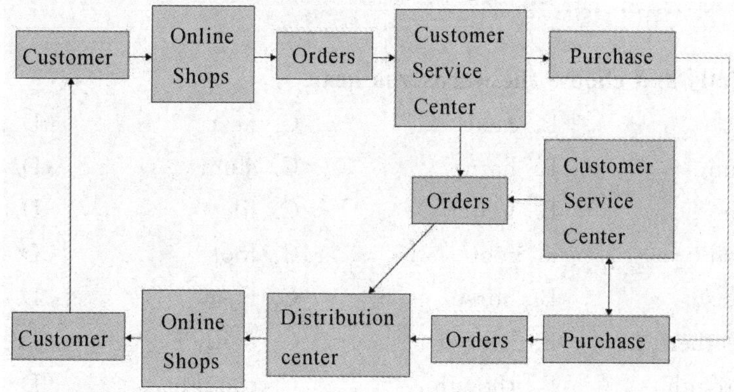

3. Suppose you are a consumer. Give a presentation to each picture in English.

Picture 1

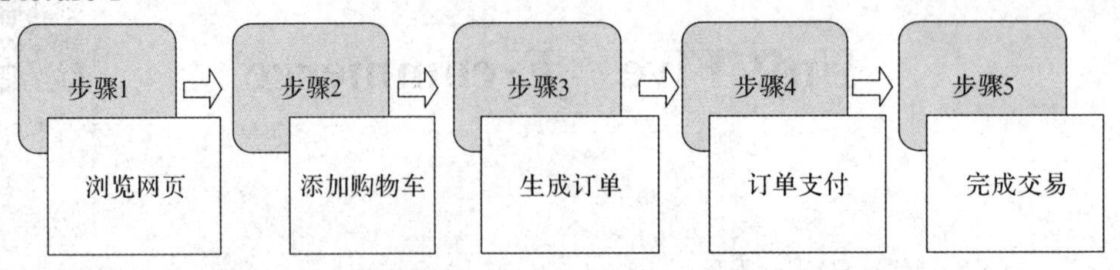

How can I get my gadgets online

Picture 2

Hints: Group buying, also known as collective buying, offers products and services at significantly reduced prices on the condition that a minimum number of buyers would make the purchase. Origins of group buying can be traced to China where it is known as Tuan Gou (Chinese: 团购) or team buying.

Section Two Listening

Task Ⅱ

1. Listen carefully and choose the words you hear.

(1) A. net B. next C. nest D. nut

(2) A. plum B. harm C. slum D. mum

(3) A. five B. knife C. file D. five

(4) A. food B. foot C. fool D. fold

(5) A. house B. horse C. hose D. host

(6) A. weather B. leather C. neither D. father

(7) A. through B. though C. thought D. throw

(8) A. bench B. beach C. bees D. bent

(9) A. pretty B. penny C. plenty D. pity

(10) A. silver B. sliver C. slaver D. slaughter

2. Listen to the short passage and choose the proper words to fill in the blanks.

There's No Better Way to Get in the Game!

High Definition Television is the best way to watch ___1___ (basketball, sports, football). And now you can get virtual ___2___ (from, flight, front) row seats thanks to Haier and the NBA. Haier, one of the world's ___3___ (famous, leading, outstanding) manufacturers of HDTVs, is now the "Official HDTV of the NBA" as well as an "Official Marketing Partner" of the NBA.

Haier is always ready to jump on the ___4___ (count, court , course,). Whether playing the game for thousands of fans or creating the technology to broadcast it to millions, both Haier and the NBA agree that high definition is ideal for NBA games. So visit this ___5___ (net, site, web) often, get in on the action, and watch for your chance to win!

3. Listen to the dialogues and choose the best responses to what you hear.

(1) A. 50 dollars.
B. 60 kg.
C. 15 feet.
D. 16 pints. ________

(2) A. Do you like my shoes?
B. Oh, the mall is so big.
C. No, thanks. I'm just hanging around.
D. What is your idea? ________

(3) A. What size does your mother wear?
B. Can you change it?
C. Where is your mother?
D. How about you? ________

(4) A. The apples are very good.
B. You like apples, don't you?
C. Yes, sir. They are over there.
D. I will tell you they are cheap. ________

(5) A. It's like this.
B. It's behind you.
C. Can I pay by credit card?
D. Where shall I go? ________

(6) A. How is this?
B. Do you want to buy something here?
C. Sure. Go ahead.

D. No, you are wrong. ________

(7) A. Okay, here you are.

B. Thank you.

C. You are welcome.

D. They look very nice. ________

(8) A. I'll take them.

B. All right. I'll take it.

C. How much is it?

D. Yes, I'd like to buy a book. ________

4. Listen to the dialogues and choose the best answer to each question you hear.

(1) A. Yes, she does. B. No, she can't.

C. Yes, she can. D. No, she doesn't.

(2) A. By credit card. B. By check.

C. In cash. D. By visa card.

(3) A. A bet. B. A pair of shoes.

C. A handbag. D. A necklace.

(4) A. A can get the newspapers. B. A can't read the English newspapers.

C. B doesn't have any newspapers. D. B doesn't know English.

(5) A. Large. B. Small.

C. Extra large. D. Medium.

5. Listen to the dialogue and answer the following questions by filling in the blanks.

(1) Where did the dialogue take place?

It's ____________.

(2) How much was the thing at discount?

It's ____________

(3) Did he want to buy it?

____

(4) Was credit card accepted in the store?

____

(5) How did he pay for it?

By ____________.

6. Listen to the passage and fill in the blanks with words you hear. Some new words given below will be of some help to you.

standard	['stændəd]n.	标准，水平
enthusiasm	[in'θjuiziæzəm] n.	热心，热情，热忱
induce	[in'djuis] v.	引诱，劝
superiority	[sjuˌpiəri'ɔriti] n.	优越，优势
handcart	['hændkɑ：t] n.	手推车
aisle	[ail] n.	通道

bumper	[ˈbʌmpə] adj.	丰盛的
naturally	[ˈnætʃərəli] adv.	自然地
environment	[inˈvaiərənmənt] n.	环境
air-condition	[aiəkənˈdiʃn] n.	空调
a variety of		种种

Supermarket—Convenient Shopping Area

Nowadays, people's living standard has been enormously __1__, and with the improvement of people's living standard, supermarkets are __2__ like mushrooms. Every day, especially at weekends, crowds of shoppers __3__ towards supermarkets and smilingly come out with rich harvests. The enthusiasm of the shoppers is induced by the superiorities of the supermarket.

Firstly, it is very convenient to do shopping in the supermarket. What you need to do is to walk your handcart along the aisles and take anything you need from the __4__ to fill your cart.

Secondly, the market __5__ a bumper variety of goods. Food, clothes, daily articles, drinks, books, home electric appliances, all are within your arm's __6__.

Thirdly, the environment is clean and __7__ with all-year-round air-condition.

Fourthly, payment is made easily, either in __8__ or with a card. And you don't have to wait long.

Naturally, people like shopping in supermarkets even they sometimes buy a few __9__ things. But the next time they do shopping, they completely __10__ it.

Section Three Reading

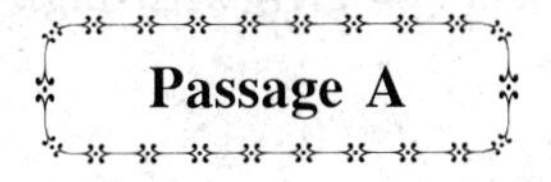

E-commerce

E-commerce (Electronic Commerce or EC) is buying and selling of goods and services on the internet, without face-to-face meeting between the two parties of the transaction.

As is known to all, the customer, the enterprise and the government are the three major entities participating in the electronic transactions or business processes. According to it, E-commerce can be generally broken into five categories: B2B, B2C, C2C, B2G and G2G. B2B, B2C and C2C are the three commonest models.

B2B (Business-to-Business)

B2B involves the transaction of a product or service from one business to another rather than to an end user. This transaction could happen between enterprises with their supply chain members or any other enterprises. For example, a furniture manufacturer requires raw materials such as wood, paint, and varnish. In B2B electronic commerce, manufactures electronically place orders with suppliers and many times payment is made electronically.

B2C (Business-to-Customer)

B2C E-commerce is a process for selling products directly to consumers from a website. In this model, enterprise and the individual customer are two main participants. Consumers browse product information pages on your website, select products and pay for them before delivery at a checkout, using a credit or debit card, or other electronic payment mechanism. Of course, consumers need to enter their address details and select one of the delivery options you offer. The basic B2C business system is relatively simple. You need a method of displaying products and prices on your website, a mechanism for recording customer details, and a checkout to accept payment.

C2C (Customer-to-Customer)

C2C is consumer to consumer. It has been the fastest growing segment of E-commerce since the advent of social networking. There are many sites offering free classifieds, auctions, and forums where individuals can buy and sell. E-Bay's auction service is a great example of where customer-to-customer transactions take place every day since 1995. At the meantime, thanks to online payment systems like PayPal, people can send and receive money online with ease.

B2G (Business-to-Government)

In B2G model, government plays the role of electronic commerce users to gain purchases from enterprise, and meanwhile performs its function of macro management to support and guide the electronic commerce.

G2G (Government-to-Government)

G2G involves E-commerce activities within a nation's government. In this model, most of the government activities can be carried out by using web technologies with high speed, improved efficiency and reduced cost.

★ *New Words*

E-commerce [iːkɒmərs] n. 电子商务
face-to-face [ˈfeɪsˌtʊˈfeɪs] adj. 面对面的 adv. 面对面地
party [ˈpɑːtɪ] n. 政党，党派；聚会，派对；当事人
transaction [trænˈzækʃ(ə)n] n. 交易；事务
customer [ˈkʌstəmə] n. 顾客
enterprise [ˈentəpraɪz] n. 企业；事业
government [ˈgʌvənmənt] n. 政府
entity [ˈentɪtɪ] n. 实体
participate [pɑːˈtɪsɪpeɪt] vi. 参与，参加
model[ˈmɒdl] n. 模型；典型；模特儿；样式

involve[ɪn'vɒlv] vt. 包含；牵涉；使陷于
raw[rɔ:] adj. 生的；未加工的
paint [peɪnt] n. 油漆；颜料，涂料
varnish ['vɑ:nɪʃ] n. 亮光漆，清漆
electronically [ˌɪlek'trɒnɪklɪ] adv. 电子地
supplier [sə'plaɪə] n. 供应厂商
payment ['peɪm(ə)nt] n. 付款，支付
website ['websaɪt] n. 网站
browse [braʊz] vt. 浏览
delivery [dɪ'lɪv(ə)rɪ] n. [贸易]交付
checkout['tʃekaʊt] n. 检验；签出；结账台
select[sɪ'lekt] v. 挑选；选拔
mechanism ['mek(ə)nɪz(ə)m] n. 机制；原理，途径
options ['ɒpʃnz] n. 选择
display [dɪ'spleɪ] n. 显示；炫耀
advent['ædvɛnt] n. 到来；出现
classifieds ['klæsifaidz] n. 分类广告
auction ['ɔ:kʃ(ə)n] vt. 拍卖；竞卖 n. 拍卖
forum ['fɔ:rəm] n. 论坛，讨论会
purchase ['pɜ：tʃəs] n. /v. 购买
macro ['mækrəʊ] adj. 巨大的，大量的 n. 宏，巨(计算机术语)
management['mænɪdʒm(ə)nt] n. 管理；管理人员；管理部门
improved [ɪm'prʊvd] adj. 改良的；改进过的
efficiency [ɪ'fɪʃ(ə)nsɪ] n. 效率；效能
reduced [rɪ'dju:st] adj. 减少的

★ *Phrases and Expressions*

on the internet	在网上
as is known to all	众所周知
participate in	参加；分享
break into	闯入；破门而入
B2B(Business to Business)	企业间电子商务模式
B2C(Business to Customer)	企业对消费者电子商务
C2C(Consumer To Consumer)	消费者间电子商务
B2G(Business to Government)	企业与政府机构间电子商务
G2G(Government to Government)	政府间的电子商务模式
end user	最终用户
supply chain	供应链
rather than	而不是

raw materials	原材料
place orders	下订单
before delivery	交货前
a credit or debit card	信用卡或借记卡
take place	发生
at the meantime	同时
with ease	熟练地，不费力地

Task Ⅲ-1

1. Fill in the blanks without referring to the passage.

As is known to all, the ________(1), the ________(2) and the ________(3) are the three major entities participating in the electronic transactions or business ________(4). According to it, E-commerce, can be generally broken into five categories: ________(5), B2C, ________(6), B2G and G2G. B2B, B2C and C2C are the three commonest models.

2. Answer the following questions according to the passage.

(1) What is E-commerce?

____

(2) What are the main types of E-commerce?

____

(3) What are the differences between B2B and B2C?

____

(4) Can you give an example of C2C model?

____

(5) What kind of role do you think the government should play in E-commerce?

____

3. Complete each of the following statements according to the passage.

(1) E-commerce is buying and selling of goods and ________ on the internet.

(2) E-commerce can be generally broken into five categories: ________, B2C, C2C, B2G and G2G.

(3) B2B involves the transaction of a product or service from one ________ to another.

(4) B2C e-commerce is a process for selling products directly to ________ from a website.

(5) ________ has been the fastest growing segment of E-commerce since the advent of social networking.

(6) In B2G model, government plays the role of ________ to gain purchases from enterprise.

4. Translate the following sentences into English.

(1) 众所周知，假冒伪劣商品损害了消费者的利益。(as is known to all)

____

(2) 我们一直都在供应链管理中平衡这两者。(supply chain)

____

(3) 我宁可乘最慢的火车去也不愿坐飞机去。(rather than)

____

(4) 由于价格太贵，我们没向这家公司订货。(place order)

____

(5) 信用证应该在交货前 15 天到 20 天由买方开出。(before delivery)

____

(6) 同时，企业也应该更为主动地承担社会责任。(at the meantime)

____

(7) 他轻而易举地通过了考试。(with ease)

____

(8) 你必须执行我的命令。(carry out)

____

5. Translate the following sentences into Chinese.

(1) E-commerce is buying and selling of goods and services on the internet, without face-to-face meeting between the two parties of the transaction.

____

(2) As is known to all, the customer, the enterprise and the government are the three major entities participating in the electronic transactions or business processes.

____

(3) B2B involves the transaction of a product or service from one business to another rather than to an end user.

____

(4) This transaction could happen between enterprises with their supply chain members or any other enterprises.

____

(5) B2C E-commerce is a process for selling products directly to consumers from a website.

____

(6) Consumers browse product information pages on your website, select products and pay for them before delivery at a checkout.

____

(7) It has been the fastest growing segment of E-commerce since the advent of social networking.

____

(8) At the meantime, thanks to online payment systems like PayPal, people can send and receive money online with ease.

____

(9) In B2G model, government plays the role of electronic commerce users to gain purchases from enterprise.

____

(10) In this model, most of the government activities can be carried out by using web technologies with high speed, improved efficiency and reduced cost.

____

How to Optimize an E-commerce Website

The world of E-commerce is constantly changing and it is harder and harder to attract visitors. There are numerous service providers online who all fight for the consumers' attention. The question is how to be visible online and reach the potentials?

The SEO professionals and consultants across the world give attention to four crucial areas while optimizing the E-commerce websites to increase the sales.

Product descriptions

Product descriptions should not only be relevant but also detailed. The potential customers read the product descriptions first, and after understanding the product or service, if they find it useful and interesting, they decide to purchase it.

The product descriptions should be also unique. If you use any content from another site, it will be considered as a duplicate content by Google, and your site may get penalized for it.

Long-tail keywords

Long-tail keywords are key words or key phrases that are more specific and usually longer-than more commonly searched for key words. Long-tail keywords get less search traffic, but will usually have a higher conversion value, as they are more specific.

Take this example: if you're a company that sells classic furniture, the chances are that your pages are never going to appear near the top of an organic search for "furniture" because there's too much competition. But if you specialize in, say, contemporary art-deco furniture, then keywords like "contemporary Art Deco-influenced semi-circle lounge" are going to reliably find those consumers looking for exactly that product.

Product reviews

About 70% of buyers search product information before making a purchase. Consumer trust other people's opinions and feedback. Make sure you give your consumer possibility to write reviews. In this way the potential buyers get objective feedback about the product quality without leaving your page. The extra benefit of product reviews is its free content that boosts your SEO value by offering fresh and unique content.

Structured data and rich snippets

Structured data is a few extra rows in the search results that are coded on URL address. Different structure will make the search result more attractive and distinguish it from the competitors.

The e-commerce rich snippet include: the stars, rating, review numbers, image and price range. The purpose of rich snippets is to provide a better review of the link content. This enables the user to make a decision before entering the page — whether the result meets the query they made.

E-commerce tracking

Rising visitor numbers doesn't mean that you can throw the gun into the bushes and rest on laurels. To understand what keywords bring real benefit and sales revenue you need to follow customers' sales path. Google Analytics which can monitor e-stores is a good choice.

★ ***New Words***

constantly ['kɒnst(ə)ntlɪ] adv. 不断地；时常地
numerous ['nju:m(ə)rəs] adj. 许多的，很多的
attention [ə'tenʃ(ə)n] n. 注意力；关心
potential [pə'tenʃl] n. 潜能；可能性 adj. 潜在的
professional [prə'feʃ(ə)n(ə)l] adj. 专业的；职业的 n. 专业人员
consultant [kən'sʌlt(ə)nt] n. 顾问；咨询者
crucial ['kru:ʃ(ə)l] adj. 重要的；决定性的
optimize ['ɒptɪmaɪz] vt. 使最优化，使完善 vi. 优化
description [dɪ'skrɪpʃ(ə)n] n. 描述，描写；类型；说明书
detailed ['di:teɪld] adj. 详细的，精细的
unique [ju:'ni:k] adj. 独特的，稀罕的 n. 独一无二的人或物
purchase ['pɜ: tʃəs] n. 购买；紧握 vt. 购买；赢得
duplicate['dju:plɪkeɪt] v. 复制 n. 副本；复制品 adj. 复制的
penalize['pi: nəlaɪz] vt. 处罚；处刑；使不利
tail[teɪl] n. 尾巴；踪迹
commonly['kɒmənlɪ] adv. 一般地；通常地
conversion[kən'vɜ:ʃ(ə)n] n. 转换；变换
classic['klæsɪk] adj. 经典的；古典的
organic [ɔ:'gænɪk] adj. 组织的；器官的；根本的
competition [ˌkɒmpə'tɪʃn] n. 竞争；比赛
contemporary [kən'temp(ə)r(ər)ɪ] adj. 当代的
art-deco [ˌɑ: t'deikəu] n. 装饰艺术
influenced ['ɪnflʊənsd] adj. 受影响的

semi-circle [sɛmɪˌsɜ:kəl] n. 半圆
lounge[laʊn(d)ʒ] n. 休息室；闲逛；躺椅
reliably [ri'laiəbli] adv. 可靠地；确实地
review [rɪ'vju:] n. 回顾；复习；评论
trust[trʌst] n. 信任，信赖；责任 v. 信任，信赖
feedback ['fi:dbæk] n. 反馈；成果，资料；回复
possibility [ˌpɒsɪ'bɪlɪtɪ] n. 可能性；可能发生的事物
quality ['kwɒlətɪ] n. 质量，[统计] 品质；特性
objective [əb'dʒektɪv] adj. 客观的；目标的 n. 目的；目标
boost [bu:st] v. 促进；增加；宣扬；推动
structured['strʌktʃəd] adj. 有结构的；有组织的
snippet ['snɪpɪt] n. 小片；片断
attractive [ə'træktɪv] adj. 吸引人的
distinguish[dɪ'stɪŋgwɪʃ] v. 区分；辨别
competitor [kəm'petɪtə] n. 竞争者，对手
rating ['reɪtɪŋ]n. 等级；等级评定
decision[dɪ'sɪʒ(ə)n] n. 决定，决心；决议
query ['kwɪərɪ] n. 疑问，质问；疑问号 ；[计] 查询 v. 询问
bush [bʊʃ] n. 灌木；矮树丛
laurel ['lɒr(ə)l] n. 月桂属植物，月桂
track [træk] n. 轨道；足迹 v. 追踪
revenue ['revənju:] n. 税收，国家的收入；收益
analytics [ænə'lɪtɪks] n. [化学][数] 分析学；解析学
choice [tʃɒɪs] n. 选择；选择权；精选品

★ ***Phrases and Expressions***

harder and harder	越来越难
service providers	服务供应商
fight for	为……而战，而奋斗
SEO(Search Engine Optimization)	搜索引擎优化
give attention to	考虑，注意，关心
potential customers	潜在客户
be considered as	被认为……
long-tail keywords	长尾关键词
search traffic	搜索流量
conversion value	转换价值
take this example	举个例子来说
specialize in	专营，专门研究，专攻

product reviews	产品评论
making a purchase	购物
structured data	结构化数据
rich snippet	富摘要
URL(Uniform Resource Locator)	全球资源定位器
distinguish... from	与……区别开
enable... to	使……能够做……
make a decision	做决定
rest on laurels	吃老本
sales revenue	产品销售收入
sales path	销售路径

Task Ⅲ-2

1. Match the following English phrases in Column A with their Chinese equivalents in Column B.

A	B
(1) SEO	a. 潜在购买者
(2) product descriptions	b. 产品评论
(3) duplicate content	c. 结构化资料
(4) Long-tail keywords	d. 全球资源定位器
(5) conversion value	e. 电子商务跟踪
(6) product reviews	f. 产品描述
(7) potential buyers	g. 长尾关键词
(8) structured data	h. 搜索引擎优化
(9) URL	i. 复制内容
(10) E-commerce tracking	j. 转化价值

2. Decide whether the following statements are T(true) or F (false) according to the passage.

(1) A good website plays a very important role in E-commerce. ( )

(2) Product descriptions should not only be relevant but also detailed and unique. ( )

(3) If you use any content from another site, it will be regarded as a duplicate content by Google but your site won't get penalized for it. ( )

(4) Long-tail keywords are key words or key phrases that are more specific and longer than more commonly searched for key words. ( )

(5) Long-tail keywords get more search traffic, and will usually have a higher conversion value. ( )

(6) To sell classic furniture, "furniture" is a better product description than "contemporary Art Deco-influenced semi-circle lounge" in an E-commerce platform. ( )

(7) Consumer trust other people's opinions and feedback. (　　)

(8) Different structure will make the search result more attractive and distinguish it from the competitors. (　　)

(9) Rich snippet is not necessary to provide a better review of the link content. (　　)

(10) Rising visitor numbers mean that you can throw the gun into the bushes and rest on laurels. (　　)

3. Translate the following passage into Chinese.

About 70% of buyers search product information before making a purchase. Consumer trust other people's opinions and feedback. Make sure you give your consumer possibility to write reviews. In this way the potential buyers get objective feedback about the product quality without leaving your page. The extra benefit of product reviews is its free content that boosts your SEO value by offering fresh and unique content.

Read the following passage and choose the best answer to each question.

Amazon. com

Amazon. com, Inc. (/ˈæməzɒn/) is an American electronic commerce and cloud computing company based in Seattle, Washington that was founded on July 5, 1994 by Jeff Bezos. Bezos selected the name Amazon by looking through the dictionary; he settled on "Amazon" because it was a place that was "exotic and different", just as he had envisioned for his Internet enterprise. The Amazon River, he noted, was the biggest river in the world, and he planned to make his store the biggest in the world. The tech giant is the largest Internet-based retailer in the world by total sales and market capitalization.

Amazon. com started as an online bookstore and later diversified to sell DVDs, Blu-rays, CDs, video downloads/streaming, MP3 downloads/streaming, audio book downloads/streaming, software, video games, electronics, apparel, furniture, food, toys, and jewelry. The company also produces consumer electronics—notably, Kindle e-readers, Fire tablets, Fire TV, and Echo—and is the world's largest provider of cloud infrastructure services (IaaS and PaaS). Amazon also sells certain low-end products like USB cables under its in-house brand Amazon Basics.

Amazon has separate retail websites for the United States, the United Kingdom and Ireland, France, Canada, Germany, Italy, Spain, Netherlands, Australia, Brazil, Japan, China, India, and Mexico. Amazon also offers international shipping to certain other countries for some of its products. In 2016, Dutch, Polish, and Turkish language versions of the German Amazon website were launched.

In 2015, Amazon surpassed Walmart as the most valuable retailer in the United States by market capitalization. Amazon is the fourth most valuable public company in the world, the largest Internet company by revenue in the world and the eighth largest employer in the United States. In 2017, Amazon announced their plans to acquire Whole Foods Market for $13.4 billion by the end of the year, which would vastly increase Amazon's presence as a physical retailer. The acquisition was interpreted by some as a direct attempt to challenge Walmart as a physical store.

In 2011, Amazon had 30,000 full-time employees in the USA, and by the end of 2016, it had 180,000 employees. The company employs 306,800 people worldwide in full and part-time jobs.

1. Which of the following statements is False according to the passage?
   A. Amazon was founded on July 5, 1994 by Jeff Bezos.
   B. Amazon is based in Seattle, Washington.
   C. Amazon.com, Inc. is an American electronic commerce and cloud computing company.
   D. Amazon is the second largest Internet-based retailer in the world by total sales and market capitalization.
2. What kind of store did Amazon.com start as?
   A. Physical bookstore.　　B. Online bookstore.
   C. CD store.　　D. Cloud infrastructure services provider.
3. Amazon separate retail websites were launched in these countries except ________.
   A. the United Kingdom　　B. Germany
   C. China　　D. Russia
4. What does the underlined word "surpassed" in Para. 4 probably mean according to the passage?
   A. 超过.　　B. 优于.　　C. 赶上.　　D. 落后.
5. What can we learn from the passage?
   A. Walmart is the most valuable retailer in the United States by market capitalization.
   B. Amazon employs the most employers in the world.
   C. Maybe one day we can buy what we need in Amazon physical stores.
   D. Nowadays Amazon is the biggest competitor of Walmart as a physical store.

Section Four　Writing

邀请信及回函(Invitation Letter and Reply)

1. 邀请信(Invitation Letter)

邀请信是指在婚礼、学术、宴会等活动中邀请有关人员时所写的信函。格式与一般书信相同。邀请信不必像请柬那么正规，但内容要更明确具体。

(1) 说明邀请对方参加什么活动、邀请的原因是什么。

(2) 将活动安排的细节及注意事项告诉对方。诸如时间、地点、参加人员、人数，做些什么样的准备及所穿的服饰等。

(3) 为了方便安排活动，如有必要，可注明请对方予以回复能否应邀及还有哪些要求等。

(4) 注意文化背景。若邀请已婚男女参加某项活动，应邀请夫妇双方共同参加。

Month date, year

Dear ________,

I will be holding ________ at ________ on ________ in order to ________. As you are a close friend of our family, my parents and I would very much like you to join us in ________ and share our joy.

The occasion will start at ________, and activities include ________, ________ and ________. In addition, there will be ________. I am sure you will enjoy a good time.

My family would feel honored by your presence.

Yours sincerely,

________(签名)

◆常用语句◆

① It is my great pleasure to invite you to our party from 3 p. m. to 5 p. m. on Saturday, October 7th.（我荣幸地邀请您[们]参加 10 月 7 日星期六下午 3 点至 5 点的聚会。）

② Some details about this activity are as follows.（活动细节如下。）

③ Please accept our warm welcome and sincere invitation.（请接受我们热情的欢迎和诚挚的邀请。）

④ Please let us know as soon as possible if you can come and tell us when you can make the trip.（请您尽早与我们联系，并告知能否成行及具体出发时间。）

⑤ We would appreciate your confirmation by October 5.（请在 10 月 5 日前给予确认，不胜感谢。）

2. 接受邀请信(Reply)

Month date, year

Dear ________,

Thank you for your kind invitation to ________ you are giving us ________. I will be very happy to come, and look forward with pleasure to meeting you.

Yours sincerely,

________(签名)

3. 婉拒邀请信(Reply)

<table><tr><td>
Month date, year
Dear ________,
　　I was grateful to receive your invitation to attend ________. Unfortunately, I have a previous ________ on that day and simply cannot rearrange my schedule at this time.
　　Best regards to you and your family.
Yours sincerely,
________(签名)
</td></tr></table>

◆常用语句◆

① Thank you for your kind invitation to your college to give lectures on"Big Data". (非常感谢您盛情邀请我到贵校做大数据的演讲。)

② I accept with pleasure the kind invitation of Mr. and Mrs. White to dinner. (我非常高兴地接受怀特夫妇的晚餐邀请。)

③ I will be very happy to come, and look forward with pleasure to meeting you. (我乐于前往，并期待与您会面。)

④ I regret that I have another engagement on that day and will not be able to attend. (我届时另外有约，故不能出席。敬请谅解。)

⑤ Please accept my sincere regrets for not being able to join your party.
(我不能参加你的派对，请接受我真诚的歉意。)

Task Ⅳ

1. You are required to write a letter according to the following instructions given in Chinese.

给 Monica 写一封短信。内容：本周六下午四点你(Paul)将和 Mary 在家里举行婚礼。真心希望最好的朋友 Monica 来参加。之后，会有小型酒会。

注意：必须包括对收信人的称谓，写信日期，发信人的签名等基本格式。

2. You are required to write both A Letter of Invitation and A Reply to the Letter according to the following instructions given in Chinese.

Letter 1

发信人：张敏

内容：(1) 获悉 Prof. Smith 要来北京外国语大学访问。

　　(2) 邀请其游览长城、故宫、颐和园等。

　　(3) 请回信告知是否能来。

写信日期：2017 年 6 月 25 日。

Letter 2

回信人:Prof. Smith

内容:(1) 感谢张敏的热情好客并接受邀请。

(2) 决定 7 月中旬来北京。

(3) 3 天正式访问结束后即可成行。

回信日期:2017 年 6 月 29 日。

Words for Reference:

长城 the Great Wall

故宫 the Imperial Palace

颐和园 the Summer Palace

Unit Six Computer Security

Unit 6

Section One Speaking

Task Ⅰ

1. **Look at the following picture and speak out each name of the following devices in computer security management in Chinese.**

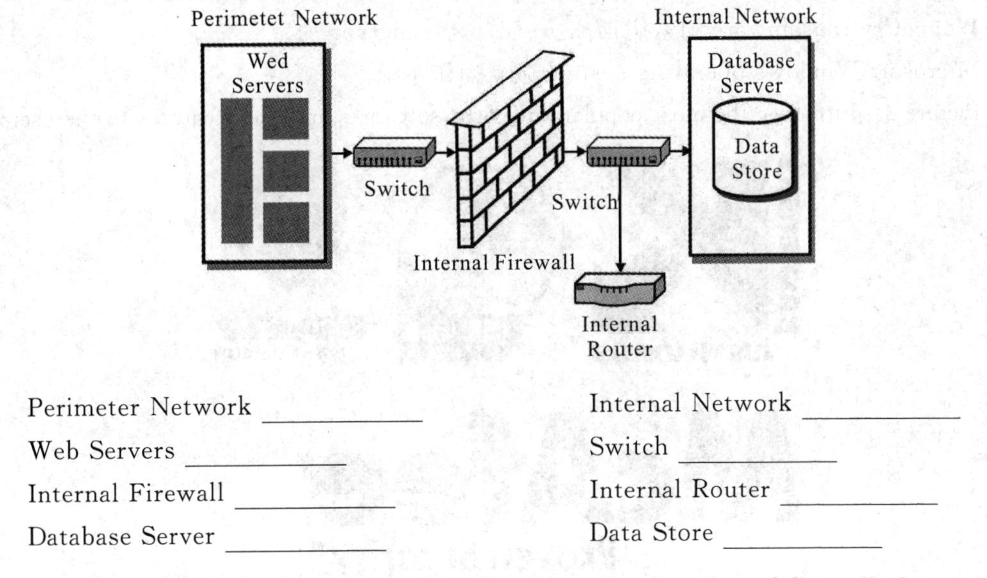

Perimeter Network ____________ Internal Network ____________

Web Servers ____________ Switch ____________

Internal Firewall ____________ Internal Router ____________

Database Server ____________ Data Store ____________

2. **Look at the following picture and describe the major function of firewalls in computer security.**

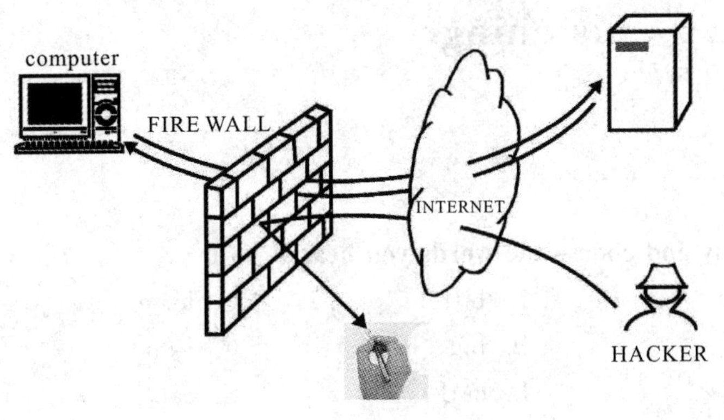

3. Suppose you are a cyber-security expert. Give a presentation to each picture in English.

Picture 1　Have you ever been a victim of malware? What have you done?

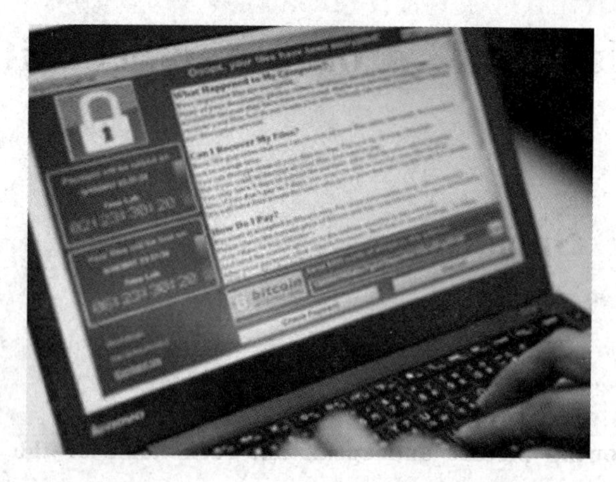

Reference words:

virus 病毒　　　　cyber attack 网络攻击

WannaCry ransomware 勒索病毒　　　　encrypt 加密

Microsoft Windows operating system 微软操作系统

Picture 2　Introduce the most popular anti-virus softwares and their features to the users.

腾讯电脑管家
GUANJIA.QQ.COM

Section Two　Listening

Task Ⅱ

1. Listen carefully and choose the words you hear.

(1) A. better　　B. bitter　　C. letter　　D. litter

(2) A. pig　　B. big　　C. pear　　D. bear

(3) A. cup　　B. cuff　　C. cap　　D. cab

(4) A. slip B. sheep C. ship D. sleep
(5) A. cheek B. chick C. copy D. coffee
(6) A. boat B. vote C. best D. vest
(7) A. hat B. hot C. vat D. vet
(8) A. beg B. bad C. bed D. bag
(9) A. seat B. seed C. seal D. zeal
(10) A. cork B. cock C. cot D. cart

2. Listen to the short passage and choose the proper words to fill in the blanks.

Not all people like to work but everyone likes to play. All over the world men and women, boys and girls enjoy __1__ (sports, shopping, sleeping). Since long, adults and children have called their friends together to spend hours, even days playing games.

Sports help people to live happily. They help to keep people healthy and feeling __2__ (wonderful , great, good). When they are playing games, people move a lot. This is good for their health. Having fun with their friends makes them happy.

Many people enjoy sports by __3__ (meeting , seeing , watching) others play. In small towns, crowds meet to watch the bicycle races or the soccer games. In the big cities, thousands buy tickets to see ice-skating shows or __4__ (football, baseball, volleyball) games.

Is it hot where you live? Then swimming is probably your __5__ (best, favorite, like) sport. Boys and girls in China love to swim, while aged people like various kinds of Tai Ji and Qi Gong. The climate is good for all the seasonal sports in the country.

3. Listen to the dialogues and choose the best responses to what you hear.

(1) A. We both do.
B. He doesn't play football.
C. No, he doesn't.
D. Yes, I do. ________

(2) A. I' m busy.
B. I usually watch TV.
C. I agree with you.
D. I don't want to go there. ________

(3) A. I don't like swimming.
B. I would like to learn it.
C. Yes, I think so.
D. It's my pleasure. ________

(4) A. Yes, he is nice.
B. Thank you.
C. He always wears a cap.
D. He is very strong. ________

(5) A. I must go now.

B. I like basketball best.

C. Three years ago.

D. one hour.

(6) A. Nobody.

B. What are you saying?

C. Well, I feel a little tired.

D. No, I don't like to.

(7) A. Yes. That's a good idea.

B. I would like a cup of tea.

C. How are you getting along?

D. Thank you for your books.

(8) A. What a pleasure.

B. Basketball.

C. Let's go.

D. I enjoy it.

4. Listen to the dialogues and choose the best answer to each question you hear.

(1) A. Very happy. B. Very funny.

C. Very excited. D. He didn't enjoy himself at all.

(2) A. He should keep this habit. B. It's good that he still does so.

C. He should change this habit. D. He'd better give up this habit.

(3) A. She likes doing morning exercises. B. She dislikes doing such exercises.

C. She can't get up early. D. She will do it.

(4) A. 2 hours. B. 3 hours.

C. Two hours and a half. D. Three hours and a half.

(5) A. Volleyball. B. Ping-pong.

C. Tennis. D. Football.

5. Listen to the dialogue and answer the following questions by filling in the blanks.

(1) What's weather like today?

It's so ____1____.

(2) What do they want to do?

They want to go ____2____.

(3) Does the woman like to do it?

____3____.

(4) Why does the man insist on playing the game?

Because they won't ____4____ when you start the game.

(5) Do they go to play the game finally?

____5____.

6. Listen to the passage and fill in the blanks with words you hear. Some new words given below will be of some help to you.

daily ['deili] adj. 每日的
swim [swim] n. 游泳
choose [tʃu: z] v. 选择
enjoyable [in'dʒɔiəb(ə)l] adj. 有趣的，愉快的
shining ['ʃainiŋ] adj. 光亮的，华丽的
breathe [bri: ð] v. 呼吸
anxiety [æŋ'zaiəti] n. 焦虑，挂念
sticky ['stiki] adj. 粘的，粘性的
enable [i'neibl] v. 使……能够
muscle ['mʌsl] n. 肌肉
figure ['figə] n. 图形，数字，形状
build up 建立
last but not least 最后但并不是最不重要的(一点)

I do like Sports!

I do like sports during my daily life. Talking about what kind of sport that I like most, I tell you that, it's swimming. The __1__ I choose swimming are as follows:

First of all, it is __2__ enjoyable, you know, in __3__. When the sun is shining __4__, and the air is even too hot to breathe, swimming would be the best __5__. Once you jump into the water, the coolness will drive all the anxiety and sticky __6__ away. Second, I guess swimming is one of the best sports to __7__ one's health. It enables your lungs to breathe more air, and helps the body to build up more muscles. Last but not least, swimming is __8__ for girls, since it's a whole body sport. __9__ the whole body's __10__, girls can gain a well-shaped figure easily.

I like sports. I like swimming.

Section Three Reading

Passage A

Computer Security

Life is full of trade offs and computer security is no different. The only safe computer is a dead computer, or at least a disconnected one. If no one can get to it, no one can harm it. The only problem is, it's not exactly useful in that state. So the extent of computer security is always a tradeoff between putting the computer to use and restricting its misuse and abuse. The time and money you spend on securing your computer has to be weighed against the likely loss if it is broken into or damaged; e. g. , you are not likely to keep your garbage under lock and key.

Computer security is defined by ISO as follows: a technological and administrative safeguard which is made or accepted for data processing system to prevent computer hardware and software from being accidentally or maliciously destroyed, alternated or exposed. The range of computer security covers physical security, software security, data security and operation security, etc.

Physical security means that the devices of computer system and its relevant facilities work regularly, including main unit, network equipments, communication circuits, storage devices, etc.

Software security refers to the integrity of software, i. e. the integrity of operating system software, database management software, network software, applications and materials concerned, including development of software, software security test, amendment and copying of software, and so on.

Data security means that data or information which is owned or created by the system is kept in full, effective and legally used without being destroyed or disclosed, including input, output, recognition of users, access control, back up and restore, and so forth.

Operation security indicates that system resources are legally used, including power management, environment(including air-conditioner), personnel, admission and exit control of rooms for computer, data and medium management, operation management, etc.

Asyou design or modify your computer and network security, think about how you want to use your systems and what you stand to lose if security is compromised. This will help guide your choice of solutions and their relative complexity and cost.

★ *New Words*

disconnected [dɪskəˈnektɪd] adj. 分离的；不连贯的；无系统的

tradeoff [ˈtredˌɔf] n. 权衡；折中

misuse [mɪsˈjuːz] vt. /n. 滥用；误用；虐待

abuse [əˈbjuːz] n. /vt. 滥用；虐待；辱骂

loss [lɒs] n. 减少；亏损；失败；遗失

garbage [ˈgɑːbɪdʒ] n. 垃圾；废物

security [sɪˈkjʊərəti] n. 安全；保证；防护

technological [ˌteknəˈlɒdʒɪkl] adj. 技术上的；工艺(学)的

administrative [ədˈmɪnɪstrətɪv] adj. 管理的，行政的

safeguard [ˈseɪfgɑːd] n. 保护，保卫；防护措施 vt. 防护；保护

accidentally [ˌæksɪˈdentəlɪ] adv. 偶然地，意外地，非故意地

maliciously [məˈlɪʃəslɪ] adv. 有敌意地

destroy [dɪˈstrɔɪ] vt. 破坏；消灭；毁坏

alternated [ɔːlˈtɜːnətid] v. 轮流(alternate 的过去分词) adj. 间隔的，轮流的

expose　[ɪk'spəʊz; ek—] vt. 揭露，揭发；使曝光；显示
integrity　[ɪn'tegrəti] n. 完整；正直，诚实；[计算机] 保存；健全
database　['deɪtəbeɪs] n. 数据库；资料库；信息库
development　[dɪ'veləpm(ə)nt] n. 发展；开发
amendment　[ə'mendmənt] n. 修正案；修改，修订
restore　[rɪ'stɔː] v. 恢复；修复
recognition　[ˌrekəg'nɪʃn] n. 认识，识别；认可；褒奖；酬劳
legally　['liːgəlɪ] adv. 法律上，合法地
personnel　[ˌpɜːsə'nel] n. 全体员工；人员；人事部门
admission　[əd'mɪʃn] n. 承认；准许进入；坦白；入场费
modify　['mɒdɪfaɪ] v. 修改，修饰；更改
compromised　['kɒmprəmaɪzd] adj. 妥协的；妥协让步的，缺乏抵抗力的
solution　[sə'lʊʃən] n. 解决方案；应对措施
relative　['relətɪv] adj. 相对的；相关的；比较而言的　n. 亲属
complexity　[kəm'pleksətɪ] n. 复杂，复杂性；复杂错综的事物

★ ***Phrases and Expressions***

weigh against	权衡，与之相当
is defined	定义
as follows	如下
prevent... from being	防止……
physical security	实体安全
main unit	主机
storage devices	存储设备
so forth	等等
trade offs	权衡，交易
stand to lose	一定失利

Task Ⅲ - 1

1. Fill in the blanks without referring to the passage.

The only safe computer is a ________(1) computer, or at least a ________(2) one. If no one can get to it, no one can harm it. The only problem is, it's not exactly ________(3) in that state. So the extent of computer security is always a ________(4) between putting the computer to use and restricting its misuse and abuse. The time and money you spend on ________(5) your computer has to be weighed against the likely loss if it is broken into or ________(6); e.g., you are not likely to keep your garbage under lock and key.

2. Answer the following questions according to the passage.

(1) Is an absolutely safe computer a perfect computer? Why?

____

(2) What is computer security?

____

(3) What does computer security mainly consist of?

____

(4) Can you describe the main content of data security?

____

(5) When you design or modify your computer and network security, what you should think about?

____

____

3. Complete each of the following statements according to the passage.

(1) The range of computer security covers physical security, ________ security, data security and operation security, etc.

(2) Physical security means that the devices of computer system and its relevant ________ work regularly.

(3) Software security refers to the ________ of software.

(4) ________ means that data or information which is owned or created by the system is kept in full, effective and legally used.

(5) Operation security indicates that ________ are legally used.

(6) The extent of computer security is always a ________ between putting the computer to use and restricting its misuse.

4. Translate the following sentences into English.

(1) 清单1展示了该表的定义。(is defined)

____

(2) 其理由如下。(as follows)

____

(3) 解锁手机，以防止不必要的检查。(prevent from)

____

(4) 病人可以食用苹果、香蕉、芒果等。(so forth)

____

(5) 他走了总有一个星期吧。(at least)

____

(6) 几个钟头抵得过整整一生，是吗？(weigh against)

____

(7) 他们闯进哪个套间了？(break into)

____

(8) 跟他作生意你是要吃亏的。(stand to lose)

____

5. Translate the following sentences into Chinese.

(1) Computer security is defined by ISO as follows: a technological and administrative safeguard which is made or accepted for data processing system to prevent computer hardware and software from being accidentally or maliciously destroyed, alternated or exposed.

____

____

(2) Physical security means that the devices of computer system and its relevant facilities work regularly.

____

____

(3) Data security means that data or information which is owned or created by the system is kept in full, effective and legally used without being destroyed.

____

____

(4) Life is full of trade offs and computer security is no different.

____

(5) The only safe computer is a dead computer, or at least a disconnected one.

____

(6) So the extent of computer security is always a tradeoff between putting the computer to use and restricting its misuse and abuse.

____

____

(7) The time and money you spend on securing your computer has to be weighed against the likely loss if it is broken into or damaged.

____

____

(8) Operation security indicates that system resources are legally used.

____

(9) As you design or modify your computer and network security, think about how you want to use your systems and what you stand to lose if security is compromised.

____

____

(10) The range of computer security covers physical security, software security, data security and operation security, etc.

____

____

Passage B

Some Security Techniques

Security is concerned with protecting information, hardware, and software. Security measures mainly consist of encryption, access control, anticipating disasters and making backup copies.

Encryption

Whenever information is sent over a network, the possibility of unauthorized access exists. The longer the distance the message has to travel, the higher the security risk is. For example, an e-mail message on a LAN meets a limited number of users operating in controlled environments such as offices. An e-mail message traveling across the country on the National Information Highway affords greater opportunities for the message to be intercepted. The government is encouraging businesses that use the National Information Highway to use a special encryption program. This program is available on a processor chip called the clipper chip and is also known as the key escrow chip. Individuals are also using encryption programs to safeguard their private communications. One of the most widely used personal encryption programs is Pretty Good Privacy.

Access Control

Security experts are constantly devising ways to protect computer systems from access by unauthorized persons. Sometimes security is a matter of putting guards on company computer rooms and checking the identification of everyone admitted. Oftentimes it is a matter of being careful about assigning passwords to people and of changing them when people leave a company.

Most major corporations today use special hardware and software called firewalls to control access to their internal computer networks. These firewalls act as a security buffer between the corporation's private network and all external networks, including the Internet. All electronic communications coming into and leaving the corporation must be evaluated by the firewall. Security is maintained by denying access to unauthorized communications.

Anticipating Disasters

Companies (and even individuals) that do not make preparations for disasters are not acting wisely. Most large organizations have a disaster recovery plan. Hardware can be kept behind locked doors, but often employees find this restriction a hindrance, so security is lax. Fire and water can do great damage to equipment. Many companies therefore will form a cooperative arrangement to share equipment with other companies in the event of catastrophe.

Backing up Data

Equipment can always be replaced. A company's data, however, may be irreplaceable. Most companies have ways of trying to keep software and data from being tampered with in the first place. They include careful screening of job applicants, guarding of passwords, and auditing of data and programs from time to time. The safest procedure, however, is to make frequent backups of data and to store them in remote locations.

★ *New Words*

encryption [ɪn'krɪpʃn] n. 加密；编密码
disaster [dɪ'zɑːstə(r)] n. 灾难；彻底的失败；不幸；祸患
unauthorized [ʌn'ɔːθəraɪzd] adj. 未经授权的；未经批准的
LAN [læn] n. 局域网
controlled [kən'trəʊld] adj. 受控制的；受约束的；克制的
afford [ə'fɔːd] vt. 给予，提供；买得起
intercept [ˌɪntə'sept] v. 拦截
encourage [ɪn'kʌrɪdʒ] vt. 鼓励，怂恿；激励；支持
identification [aɪˌdentɪfɪ'keɪʃ(ə)n] n. 鉴定，识别；认同；身份证明
password ['pɑːswɜːd] n. 密码；口令
assign [ə'saɪn] vt. 分配；指派
corporation [kɔːpə'reɪʃ(ə)n] n. 公司；法人(团体)
firewall ['faɪəwɔl] n. 防火墙
evaluate [ɪ'væljʊeɪt] v. 评价；估价
maintain [mein'tein] vt. 维持；维修；供养
deny [dɪ'naɪ] vt. 否定，否认 vi. 否认；拒绝
anticipate [æn'tɪsɪpeɪt] vt. 预期，期望；占先
recovery [rɪ'kʌv(ə)rɪ] n. 恢复，复原
restriction [rɪ'strɪkʃ(ə)n] n. 限制；约束；束缚
hindrance ['hɪndr(ə)ns] n. 障碍；妨碍
lax [læks] adj. 松的；松懈的；腹泻的
cooperative [kəʊ'ɒpərətɪv] adj. 合作的
catastrophe [kə'tæstrəfɪ] n. 大灾难；大祸；惨败
replace [rɪ'pleɪs] vt. 取代，代替
irreplaceable [ɪrɪ'pleɪsəb(ə)l] adj. 不能替代的，不能调换的
tamper ['tæmpə] vi. 篡改
screening ['skriːnɪŋ] n. 筛选；审查 v. 筛选 adj. 筛选的
applicant ['æplɪk(ə)nt] n. 申请人，申请者
audit ['ɔːdɪt] vi. 审计 n. 审计
procedure [prə'siːdʒə] n. 程序，手续；步骤
backups ['bækʌps] n. [计] 备份
remote [rɪ'məʊt] adj. 遥远的 n. 远程

★ *Phrases and Expressions*

be concerned with 涉及，参与

access control	限制访问
backup copies	副本，备份件
unauthorized access	越权存取
processor chip	处理器芯片
clipper chip	加密芯片
key escrow chip	密钥托管芯片
devise ways to	想办法
a matter of	大约，……的问题
security buffer	安全缓冲区
make preparations	做准备工作
do damage to	损坏，损害
tampered with	篡改
in the first place	首先，第一，原本
in the event of	如果，如果……发生，万一
back up	支援，备份
ways of	依靠，方法
from time to time	不时，有时

Task Ⅲ – 2

1. Match the following English phrases in Column A with their Chinese equivalents in Column B.

A	B
(1) security measures	a. 设置密码
(2) access control	b. 备份数据
(3) anticipating disasters	c. 安全措施
(4) unauthorized access	d. 越权存取
(5) backup copies	e. 加密芯片
(6) clipper chip	f. 安全缓冲区
(7) assigning passwords	g. 副本，备份
(8) computer networks	h. 限制访问
(9) security buffer	i. 预防灾难
(10) backing up data	j. 计算机网络

2. Decide whether the following statements are T(true) or F (false) according to the passage.

(1) Whenever information is sent over a network, the possibility of unauthorized access exists. (　　)

(2) The shorter the distance the message has to travel, the higher the security risk is. (　　)

(3) An e-mail message on the National Information Highway affords greater opportunities to be intercepted than on LAN. (　　)

(4) Individuals needn't use encryption programs to safeguard their private communications. ( )

(5) Oftentimes security is a matter of being careful about assigning passwords to people and of changing them when people leave a company. ( )

(6) Firewalls are often used by corporations to control access to their internal computer networks. ( )

(7) Only important electronic communications coming into and leaving the corporation must be evaluated by the firewall. ( )

(8) Companies (and even individuals) that do not make preparations for disasters are not acting wisely. ( )

(9) Fire and water can do great damage to equipment. ( )

(10) The safest procedure is to screen job applicants, guard passwords, and audit data and programs from time to time. ( )

3. Translate the following passage into Chinese.

Equipment can always be replaced. A company's data, however, may be irreplaceable. Most companies have ways of trying to keep software and data from being tampered with in the first place. They include careful screening of job applicants, guarding of passwords, and auditing of data and programs from time to time. The safest procedure, however, is to make frequent backups of data and to store them in remote locations.

Extra Reading

Read the following passage and choose the best answer to each question.

Computer Viruses

A computer virus is a program or a piece of code that is loaded on your computer without your knowledge and run against your wishes. The common viruses computers always suffer are viruses, worms and Trojans.

Virus

A computer virus may modify computer and can self-replicate. It requires a host program that executes infected files by the user.

Worm

Spreading itself automatically through computer networks by email, a computer worm doesn't request a host program to infect and manipulate systems.

Trojan

A Trojan horse is a program that appears harmless but hides malicious functions. The user unconsciously, or even consciously download and install an. exe file first.

Once a computer virus is running, it can infect other programs or documents. So how to protect your computer from viruses?

Keep the Operating System updated

The first step in protecting your computer from any malicious viruses there is to ensure that your Operating System is up-to-date. This is essential if you are running a Microsoft Windows OS.

Install an antivirus

The main scope of an antivirus is to disinfect an infected computer. Most antivirus software comes with a real-time protection where it can scan for any incoming data in real-time and worn the user when there is any suspicious activity.

Use a firewall

You should also install a firewall. A firewall is a system that prevents unauthorized use and access to your computer. It can be either hardware or software. For individual home users, the most popular firewall choice is a software firewall.

1. Virus in computer is ________.
   A. a germ　　B. a computer program
   C. a useful application　　D. a document
2. ________needs a host program to execute the infected files.
   A. A virus　　B. A worm
   C. A Trojan horse　　D. A firewall
3. ________appears harmless, but hides malicious functions.
   A. A virus　　B. A worm
   C. A Trojan horse　　D. All the above
4. According to the passage, we can do all these to protect our computer from viruses except ________.
   A. keeping the Operating System updated　　B. installing an antivirus
   C. turning off the power　　D. using a firewall
5. After reading the passage, which of the following statement do you think is true?
   A. Installing a firewall is an effective way to protect our computer.
   B. A firewall can only be software.
   C. We always choose hardware firewall at home.
   D. A firewall is a system that prevents authorized use and access to your computer.

Section Four　Writing

询价信 & 报价信(Inquiry Letter & Quotation Letter)

1. 询价信(Inquiry Letter)

在对外贸易中，询盘，也叫询价(inquiry 或 enquiry)是买方对于所要购买的商品向卖方作出的询问。

普通询盘(a general inquiry):索取普通资料，如目录(a catalogue)、价目表或报价单(a price-list or quotation sheets)、样品(a sample)、图片(illustrated photo prints)等。

具体询盘(a specific inquiry):具体询问商品名称(the name of the commodity)、规格(the specifications)、数量(the quantity)、单价(the unit price FOB.../CIF...)，装船期(the time of shipment)、付款方式(the terms of payment)等。

Month date, year

Dear ________,

We have seen your advertisement on the newspaper. We are interested in ________ made by your company. Could you please send us your current catalogue and price list?

We are looking forward to your early reply.

Yours sincerely,

________(signature)

________(position)

◆常用语句◆

① Your company was recommended to me by a colleague.(一位同事向我推荐贵公司。)

② We are interested in your product of...(我们对贵公司生产的……很感兴趣。)

③ Could you provide such details about them as prices, discounts, delivery dates, and terms of payment?(请告知有关产品的价格、折扣、发货日期及付款方式。)

④ We would appreciate it if you could send us the information requested.(如能惠寄所需材料，将感激不尽。)

2. 报价信(Quotation Letter)

报价信是商务活动中作为卖方在接到客户询价函后发出的回复性信函。

对于卖方而言，一封报价信可能意味着一次销售的好时机，所以回复的报价函一定要及时、确切、周到，不要因为某些小小疏忽而失去了潜在客户。

Month date, year

Dear ________,

Thank you for your interest in ________(产品名称) made by our company. A catalogue and a price list are attached to this letter.

Yours sincerely,

________(签名)

________(职务)

◆常用语句◆

① Thank you for your interest in the computer made by our HP company.（感谢您对我们 HP 公司生产的电脑感兴趣。）

② A catalogue and a price list are attached to this letter.（随信附寄了产品目录及价目单。）

③ A visit by your representative would be greatly appreciated.（非常欢迎贵公司派代表来我处参观。）

Task Ⅳ

1. You are required to write an Inquiry Letter according to the following information given in Chinese.

发信人：Professor Li Ming　　　　　　收信人：Maria Stein

发信日期：2017 年 3 月 29 日

内容：Professor Li Ming 欲购买 Maria Stein 所在的 HP 公司生产的激光打印机（jet printer），写信了解有关价格以及售后服务的情况。

2. You are required to write a Quotation Letter according to the following information given in Chinese to reply to the Inquiry Letter above.

发信人：Maria Stein　　　　　　收信人：Professor Li Ming

发信日期：2017 年 4 月 2 日

内容：Maria Stein 非常感谢 Professor Li Ming 对他们公司生产的激光打印机感兴趣，随信附上目录以及价目表，并告知 HP 公司为客户提供优良的服务。不仅如此，如果购买的数量较大，还可以享受折扣。

Unit Seven Big Data

Unit 7

Section One Speaking

Task I

1. Look at the following graphic and speak out some words related big data in Chinese.

search ____________ information ____________

internet ____________ storage ____________

analysis ____________ petabyte ____________

2. List the Chinese names of the following graphic.

Big Data in Computer Science

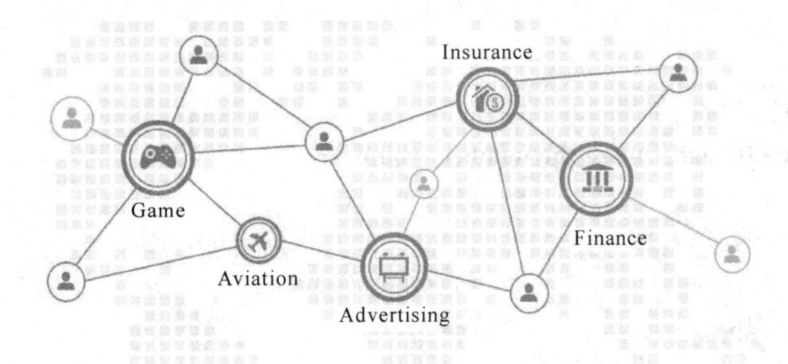

3. Suppose you are a computer engineer. Give a presentation to each picture in English.

Picture 1

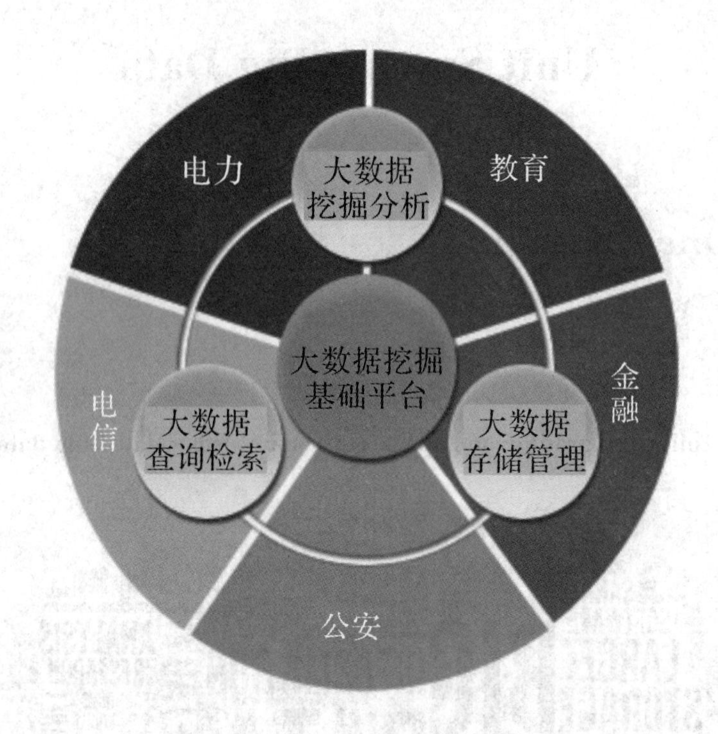

Picture 2

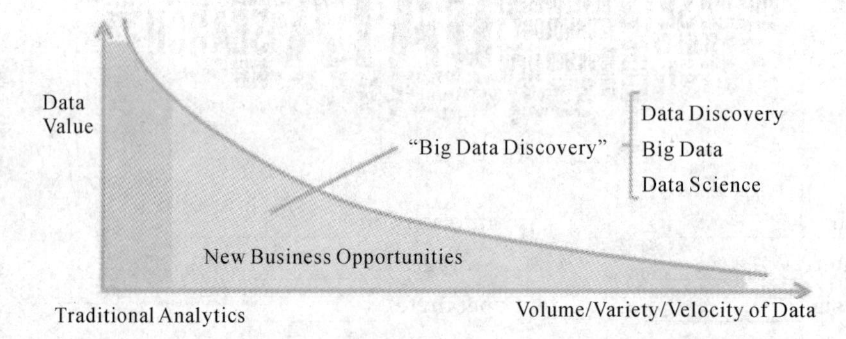

Section Two Listening

Task Ⅱ

1. Listen carefully and choose the numbers you hear.

(1) 15 / 50 (2) 140 / 114

(3) 70% / 17% (4) 20, 000, 000 / 200, 000, 000

(5) 13 / 30 (6) 78.21 / 7, 821

(7) 380 / 318 (8) 17 / 70

(9) 60% / 16% (10) 55, 000, 000, 000 / 5, 500, 000, 000

2. Listen to the short passage and choose the proper words to fill in the blanks.

The Industrial Revolution, in the 18th and 19th centuries, brought a new kind of __1__ (advertisement, advertiser, advertising). Large factories took the place of small workshops and goods were produced in large quantities. Manufactures used the newly built railroads to distribute their __2__ (products, produce, production) over wide areas. They had to find many __3__ (hundreds, thousands, thousand) of customers in order to stay in business. They could not __4__ (only, nearly, simply) tell the people where shoes or cloth or tea could be bought—they had to learn how to make people want to buy a specific product. Thus modern advertising was __5__ (born, bought, brown).

3. Listen to the questions and choose the best responses to what you hear.

(1) A. He is leaving by bus.
B. He has a big family.
C. He is living in a small town.
D. He is working as a lawyer. ______

(2) A. Sure. Here you are.
B. Yes, please give it to me.
C. Sorry, I can't help you.
D. No, I can take it myself. ______

(3) A. Yes, you can drive it.
B. Should I go with you?
C. No, everything is fine.
D. Take your time. There is no hurry. ______

(4) A. Yes, I got it.
B. It's interesting.
C. No, I don't have it.
D. It begins at 6: 00. ______

(5) A. It's difficult to park here.
B. I don't like taking it.
C. I came here by bus.
D. I agree with you. ______

(6) A. It's open at 9 a. m.
B. Sorry, I have no idea.
C. That's all right.
D. Thank you. ______

(7) A. She's an English student.
B. She's interested in music.
C. She's a friend of mine.
D. She's tall with dark hair. ______

(8) A. It's not too expensive.

B. It is still early.

C. You are so kind.

D. Go straight ahead.

4. Listen to the dialogues and choose the best answer to each question you hear.

(1) A. His bag. B. His tape.

C. His cap. D. His book.

(2) A. It was wonderful. B. It was disappointing.

C. It was boring. D. It was unusual.

(3) A. Manager and secretary. B. Doctor and patient.

C. Shop assistant and customer. D. Taxi driver and passenger.

(4) A. At 1:40. B. At 1:50.

C. At 2:00. D. At 3:50.

(5) A. He was killed in an air crash. B. He was wounded in a fight.

C. He was injured in an accident. D. He was burnt in a fire.

5. Listen to the dialogue and choose the best answer to each question you hear.

(1) Where is Mary?

A. She is in the sitting-room. B. She is in the garden.

C. She is in the bedroom. D. She is in the kitchen.

(2) What did Mary ask John to do?

A. She asked him to help her. B. To look for their baby.

C. To tell her a story. D. To do some washing.

(3) What did John answer?

A. The baby was sleeping B. He was playing with the baby.

C. The baby was upstairs. D. The baby was crying.

(4) Where is the baby actually?

A. In the sitting-room. B. In the kitchen.

C. In the washroom. D. In the garden.

(5) What is the baby doing?

A. He is playing games. B. He is watching TV.

C. He is listening to the radio. D. He is cleaning his shoes.

6. Listen to the passage and fill in the blanks with words you hear. Some new words given below will be of some help to you.

fact	[fækt]	n.	事实；现实
count	[kaunt]	v.	计数；计算
respond	[ris'pɔnd]	v.	相应；反应
advertiser	['ædvətaɪzə]	n.	刊登广告者；广告客户
gain	[gein]	v.	获得；吸引
business	['biznis]	n.	企业；行业；营业；商业

fascinating ['fæsineɪtɪŋ] adj. 迷人的；吸引人的
attraction [ə'trækʃən] n. 吸引；吸引力
excitement [ɪk'saɪtmənt] n. 激动；兴奋
literature ['lɪtərɪtʃə] n. 文学；文学作品
react [rɪ'ækt] v. 起反应；起作用
prevent [prɪ'vent] v. 防止；预防
leaf through 翻阅

Advertising

Advertising is part of our daily lives. To __1__ this fact, you have only to leaf through a magazine or newspaper or __2__ the radio or television commercials that you hear in one __3__. Most people see and hear several hundred advertising messages every day. And people __4__ to the many devices that advertisers use to gain their attention.

Advertising is a big __5__—and to many people, a fascinating business, filled with attraction and __6__. It is part literature, part art, and part show business.

Advertising is the difficult business of bringing __7__ to great number of people. The purpose of an advertisement is to make people respond—to make them __8__ to an idea, such as helping prevent __9__ fires, or to make them want to buy a certain __10__ or service.

Section Three Reading

Passage A

Big Data（Ⅰ）

Big data is a term for data sets that are so large or complex that traditional data processing application software is inadequate to deal with them. Big data challenges include capturing data, data storage, data analysis, search, sharing, transfer, visualization, querying, updating and information privacy.

Lately, the term "big data" tends to refer to the use of predictive analytics, user behavior analytics, or certain other advanced data analytics methods that extract value from data, and seldom to a particular size of data set. "There is little doubt that the quantities of data now available are indeed large, but that's not the most relevant characteristic of this new data ecosystem." Analysis of data sets can find new correlations to "spot business trends, prevent diseases, combat crime and so on." Scientists, business executives, practitioners of medicine, advertising and governments alike regularly meet difficulties with large data-sets in areas including Internet search, fintech, urban informatics, and business informatics. Scientists encounter limitations in e-Science work, including meteorology, genomics, connectomics, complex physics simulations, biology and environmental research.

Data sets grow rapidly in part because they are increasingly gathered by cheap and numerous information-sensingInternet of things devices such as mobile devices, aerial (remote sensing), software logs, cameras, microphones, radio-frequency identification (RFID) readers and wireless sensor networks. The world's technological per-capita capacity to store information has roughly doubled every 40 months since the 1980s, as of 2012, every day 2.5 exabytes (2.5×10^{18}) of data are generated. One question for large enterprises is determining who should own big-data initiatives that affect the entire organization.

Relational database management systems and desktop statistics and visualization-packages often have difficulty handling big data. The work may require "massively parallel software running on tens, hundreds, or even thousands of servers". What counts as "big data" varies depending on the capabilities of the users and their tools, and expanding capabilities make big data a moving target. "For some organizations, facing hundreds of gigabytes of data for the first time may trigger a need to reconsider data management options. For others, it may take tens or hundreds of terabytes before data size becomes a significant consideration."

★ *New Words*

inadequate [ɪn'ædɪkwət] adj. 不充分的，不适当的

challenge ['tʃælɪn(d)ʒ] n. 挑战；怀疑 vt. 向……挑战

visualization [ˌvɪzjʊəlaɪ'zeɪʃən] n. 形象化；可视性

update [ʌp'deɪt] vt. 更新；校正，修正；使现代化 n. 更新；现代化

predictive [prɪ'dɪktɪv] adj. 预言性的；成为前兆的

ecosystem ['iːkəʊsɪstəm] n. 生态系统

correlation [ˌkɒrə'leʃən] n. 统计；相互关系

executive [ɪg'zekjʊtɪv] adj. 经营的；执行的 n. 总经理；执行者

practitioner [præk'tɪʃ(ə)nə] n. 开业者，从业者

fintech ['faɪntek] n. 金融科技

informatics [ˌɪnfə'mætɪks] n. [计] 信息学；情报学(复数用作单数)

encounter [ɪn'kaʊntə] vt. 遭遇，邂逅；遇到 n. 遭遇

limitation [lɪmɪ'teɪʃ(ə)n] n. 限制；限度；极限

meteorology [ˌmiːtɪə'rɒlədʒɪ] n. 气象状态，气象学

genomics [dʒə'nɒmɪks] n. 基因组学；基因体学

connectomics ['kənektɒmɪks] n. 关系学

simulations [ˌsɪmjʊ'leɪʃən] n. [计] 模拟(simulation 的复数)；[计] 仿真

biology [baɪ'ɒlədʒɪ] n. 生物；生物学
environmental [ɪnvaɪrən'ment(ə)l] adj. 环境的，周围的；有关环境的
aerial ['eərɪəl] adj. 空中的，航空的；空气的；空想的 n. [电讯] 天线
per-capita [pə 'kæpɪtə] n. [统计] 人均；(拉丁)每人；按人口计算
exabyte [eksəˌbaɪt] n. 百亿亿字节，艾字节
generate ['dʒenəreɪt] vt. 使形成；发生
initiative [ɪ'nɪʃɪətɪv] n. 主动权 adj. 主动的；自发的；起始的
massively ['mæsivli] adv. 大量地；沉重地；庄严地
parallel ['pærəlel] n. 平行线；对比 vt. 使……与……平行 adj. 平行的
trigger ['trɪgə] vt. 引发，引起；触发 n. 扳机；[电子] 触发器
terabyte ['terəbait] n. 太字节；兆兆位(量度信息单位)
consideration [kənsɪdə'reɪʃ(ə)n] n. 考虑；原因；关心；报酬

★ ***Phrases and Expressions***

big data	大数据
data set	数据集
deal with	处理
capturing data	数据采集
spot business trend	发现市场趋势
combat crime	治理犯罪

Task Ⅲ - 1

1. Fill in the blanks without referring to the passage.

Big data is a term for ________(1) that are so large or ________(2) that traditional data processing ________(3) software is ________(4) to deal with them. Big data challenges include ________(5), data storage, data analysis, search, sharing, transfer, visualization, querying, updating and information ________(6).

2. Answer the following questions according to the passage.

(1) What is big data?

____

(2) What can analysis of data sets do?

____

(3) Why do data sets grow rapidly?

____

(4) How much does the world's technological per—capita capacity to store information have?

____

____

(5) Why do relational database management systems and desktop statistics and visualization-packages often have difficulty handling big data?

____

____

3. Complete each of the following statements according to the passage.

(1) Big data is a term for ________ that are so large or complex that traditional data processing application software is inadequate to deal with them.

(2) Analysis of data sets can find ________ to "spot business trends, prevent diseases, combat crime and so on."

(3) Data sets grow rapidly in part because they are increasingly gathered by cheap and numerous ________.

(4) One question for large enterprises is determining who should own ________ that affect the entire organization.

(5) Handling big data. may require "________ running on tens, hundreds, or even thousands of servers".

(6) For some organizations, facing hundreds of gigabytes of data for the first time may trigger a need to reconsider ________.

4. Translate the following sentences into English.

(1) 大数据将如何改变您的做事方式?(big data)

____

(2) 你知道的,这个领域的信息和知识更新得很快。(update)

____

(3) 稍后您要向该数据集添加更多的输入或者输出列。(data set)

____

(4) 语言教师常从语法书里摘录例子。(extract)

____

(5) 数字财富管理是一种金融科技门类,仍然处于上升期。(fintech)

____

(6) 我把这一切归因于他工作既勤奋又主动。(initiative)

____

(7) 我们对于所有项目都将从同一个用户数据库起步。(database)

____

(8) 你可以纠正任何可能在模型中遇到的错误。(encounter)

____

5. Translate the following sentences into Chinese.

(1) Big data is a term for data sets that are so large or complex that traditional data processing application software is inadequate to deal with them.

____

____

(2) Big data challenges include capturing data, data storage, data analysis, search, sharing, transfer, visualization, querying, updating and information privacy.

____

____

(3) Analysis of data sets can find new correlations to "spot business trends, prevent diseases, combat crime and so on."

____

____

(4) Scientists encounter limitations in e-Science work, including meteorology, genomics, connectomics, complex physics simulations, biology and environmental research.

____

____

(5) Data sets grow rapidly in part because they are increasingly gathered by cheap and numerous information-sensing Internet of things devices.

____

____

(6) The world's technological per-capita capacity to store information has roughly doubled every 40 months since the 1980s, as of 2012, every day 2.5 exabytes (2.5×10^{18}) of data are generated.

____

____

(7) One question for large enterprises is determining who should own big-data initiatives that affect the entire organization.

____

____

(8) Relational database management systems and desktop statistics and visualization-packages often have difficulty handling big data.

____

____

(9) The work may require "massively parallel software running on tens, hundreds, or even thousands of servers".

____

____

(10) What counts as "big data" varies depending on the capabilities of the users and their tools, and expanding capabilities make big data a moving target.

____

____

Big Data(Ⅱ)

Big data, the term has been in use since the 1990s. Big data usually includes data sets with sizes beyond the ability of commonly used software tools to capture, curate, manage, and process data within a tolerable elapsed time. Big Data philosophy encompasses unstructured, semi-structured and structured data, however the main focus is on unstructured data. Big data "size" is a constantly moving target, as of 2012 ranging from a few dozen terabytes to many petabytes of data. Big data requires a set of techniques and technologies with new forms of integration to reveal insights from datasets that are diverse, complex, and of a massive scale.

In a 2001 research report and related lectures, META Group (now Gartner) defined data growth challenges and opportunities as being three-dimensional, i. e. increasing volume (amount of data), velocity (speed of data in and out), and variety (range of data types and sources). Gartner, and now much of the industry, continue to use this "3Vs" model for describing big data. In 2012, Gartner updated its definition as follows: "Big Data is high-volume, high-velocity and/or high-variety information assets that demand cost-effective, innovative forms of information processing that enable enhanced insight, decision making, and process automation." Gartner's definition of the 3Vs is still widely used, and in agreement with a consensual definition that states that "Big Data represents the information assets characterized by such a High Volume, Velocity and Variety to require specific Technology and Analytical Methods for its transformation into Value". Additionally, a new V "Veracity" is added by some organizations to describe it, revisionism challenged by some industry authorities. The 3Vs have been expanded to other complementary characteristics of big data:

- Volume: big data doesn't sample; it just observes and tracks what happens;
- Velocity: big data is often available in real-time;
- Variety: big data draws from text, images, audio, video; plus it completes missing pieces through data fusion;
- Machine learning: big data often doesn't ask why and simply detects patterns;
- Digital footprint: big data is often a cost-free byproduct of digital interaction;

The growing maturity of the concept more starkly delineates the difference between big data and Business Intelligence:

- Business Intelligence uses descriptive statistics with data with high information density to measure things, detect trends, etc..
- Big data uses inductive statistics and concepts from nonlinear system identification to infer laws (regressions, nonlinear relationships, and causal effects) from large sets of

data with low information density to reveal relationships and dependencies, or to perform predictions of outcomes and behaviors.

★ ***New Words***

curate [ˈkjʊərət] v. 注意，组织
tolerable [ˈtɒl(ə)rəb(ə)l] adj. 可以的；可容忍的
elapse [ɪˈlæps] vi. 消逝；时间过去 n. 流逝
encompass [ɪnˈkʌmpəs] vt. 包含；包围，环绕；完成
unstructured [ʌnˈstrʌktʃəd] adj. 无社会组织的；松散的；非正式组成的
semi-structured [semɪˈstrʌktʃəd] adj. 半结构化的
integration [ɪntɪˈgreɪʃ(ə)n] n. 集成；综合
reveal [rɪˈviːl] vt. 显示；透露
insight [ˈɪnsaɪt] n. 洞察力；洞悉
diverse [daɪˈvɜːs; ˈdaɪvɜːs] adj. 不同的；多种多样的；变化多的
three-dimensional [ˈθridaɪˈmenʃənəl] adj. 三维的；立体的；真实的
velocity [vəˈlɒsəti] n.（物）速度
innovative [ˈɪnəvətɪv] adj. 革新的，创新的；新颖的
enhance [ɪnˈhɑːns] vt. 提高；加强；增加
consensual [kənˈsensjʊəl] adj. 交感的；在双方同意下成立的
veracity [vəˈræsəti] n. 诚实；精确性；老实
revisionism [rɪˈvɪʒənɪzəm] n. 修正主义
authority [ɔːˈθɒrɪtɪ] n. 权威；权力；当局
complementary [kɒmplɪˈment(ə)rɪ] adj. 补足的，补充的
observe [əbˈzɜːv] vt. 庆祝 vt. 观察；遵守
fusion [ˈfjuːʒ(ə)n] n. 融合；熔化；熔接；融合物
detect [dɪˈtekt] v. 发现
byproduct [ˈbaɪˌprɒdʌkt] n. 副产品
maturity [məˈtʃʊərətɪ] n. 成熟；到期；完备
starkly [ˈstaːkli] adv. 严酷地；明显地
delineate [dɪˈlɪnɪeɪt] vt. 描绘；描写；画……的轮廓
density [ˈdensɪtɪ] n. 密度
inductive [ɪnˈdʌktɪv] adj. [数] 归纳的；[电] 感应的；诱导的
nonlinear [nɒnˈlɪnɪə] adj. 非线性的
regression [rɪˈgreʃ(ə)n] n. 回归；退化；逆行；复原
dependency [dɪˈpend(ə)nsɪ] n. 属国；从属；从属物
outcome [ˈaʊtkʌm] n. 结果，结局；成果

★ ***Phrases and Expressions***

information asset 信息资产

information processing	信息加工
digital footprint	数据痕迹
business Intelligence	商业智能
inductive statistic	归纳统计

Task Ⅲ-2

1. Match the following English phrases in Column A with their Chinese equivalents in Column B.

A	B
(1) data set	a. 治理犯罪
(2) capturing data	b. 大数据
(3) big data	c. 市场趋势
(4) combat crime	d. 数据集
(5) Internet of things device	e. 信息资产
(6) business trend	f. 商业智能
(7) digital footprint	g. 数据痕迹
(8) inductive statistic	h. 数据采集
(9) Business Intelligence	i. 物联网设备
(10) information asset	j. 归纳统计

2. Decide whether the following statements are T(true) or F (false) according to the passage.

(1) Big data, the term has been in use since 1990. ( )

(2) Big data "size" is a constantly moving target, as of 2012 ranging from a few dozen terabytes to many petabytes of data. ( )

(3) Big data requires a set of techniques and technologies with new forms of integration to reveal insights from database. ( )

(4) Gartner and much of the industry continue to use this "2Vs" model for describing big data now. ( )

(5) The 3Vs have been expanded to other complementary characteristics of big data. ( )

(6) Big data does sample, it observes and tracks what happens. ( )

(7) Big data often doesn't ask why and simply detects patterns. ( )

(8) Big data isn't a cost-free byproduct of digital interaction. ( )

(9) Big data uses descriptive statistics with data with high information density to measure things, detect trends, etc.. ( )

(10) Big data uses inductive statistics and concepts from nonlinear system identification to infer laws. ( )

3. Translate the following passage into Chinese.

Big data, the term has been in use since the 1990s. Big data usually includes data sets with sizes beyond the ability of commonly used software tools to capture, curate, manage, and process data within a tolerable elapsed time. Big Data philosophy encompasses unstructured, semi-structured and structured data, however the main focus is on unstructured data. Big data "size" is a constantly moving target, as of 2012 ranging from a few dozen terabytes to many petabytes of data. Big data requires a set of techniques and technologies with new forms of integration to reveal insights from datasets that are diverse, complex, and of a massive scale.

Extra Reading

Read the following passage and choose the best answer to each question.

The Characteristics of big data

Big data can be described by the following characteristics:

Volume

The quantity of generated and stored data. The size of the data determines the value and potential insight and whether it can actually be considered big data or not.

Variety

The type and nature of the data. This helps people who analyze it to effectively use the resulting insight.

Velocity

In this context, the speed at which the data is generated and processed to meet the demands and challenges that lie in the path of growth and development.

Variability

Inconsistency of the data set can hamper processes to handle and manage it.

Veracity

The quality of captured data can vary greatly, affecting the accurate analysis.

Factory work and Cyber-physical systems may have a 6C system:

- Connection (sensor and networks)
- Cloud (computing and data on demand)
- Cyber (model and memory)
- Content/context (meaning and correlation)
- Community (sharing and collaboration)
- Customization (personalization and value)

Data must be processed with advanced tools (analytics and algorithms) to reveal meaningful information. For example, to manage a factory one must consider both visible and invisible issues with various components. Information generation algorithms must detect and address invisible issues such as machine degradation, component wear, etc. on the factory floor.

1. Which main characteristics can big data be described by?

   A. Volume. B. Variety. C. Velocity. D. All the above.

2. Which element of data determines the value and potential insight and whether it can actually be considered big data or not?

   A. The size of the data. B. The type and nature of the data.

   C. The quality of captured data. D. The inconsistency of the data.

3. Which characteristic of big data helps people analyze data to effectively use the resulting insight?

   A. Variability. B. Velocity. C. Variety. D. Veracity.

4. Which characteristic of big data affect the accurate analysis?

   A. Variability. B. Velocity. C. Variety. D. Veracity.

5. Factory work and Cyber-physical systems may have a ________ system according to the passage.

   A. 5C B. 6C C. 7C D. 8C

Section Four Writing

投诉信和求职信(Complaint Letter & Application Letter)

1. 投诉信(Complaint Letter)

写投诉信应本着真实的原则，如实地反映情况。投诉信应包括以下几部分：

(1) 投诉者的姓名、性别、国籍、职业、单位(团体)名称、地址、联系电话。

(2) 被投诉者的名称、通讯地址、联系电话。

(3) 投诉的事实与理由。

(4) 具体赔偿要求。

(5) 与事实有关的证明材料，如合同、传真、机船车票、凭证、发票等。

Month date, year

Dear ________,

I am writing to complain about the ________(产品名称). ____________ ____. I strongly insist that you replace it or refund me as soon as possible.

Yours faithfully,

________(签名)

◆常用语句◆

① I am writing to complain about the camera that I bought in your supermarket last month.（今去信就上月在贵超市购买的相机提出索赔。）

② I strongly insist that you replace it or refund me as soon as possible.（请尽快给予更换或退款。）

③ We would be obliged if you would forward us a replacement for the machine as soon as possible.（若你方尽快更换那台机器，我方将感激不尽。）

④ The quality of your shipment for our order is not in conformity with the agreed specifications, we must therefore lodge a claim against you for the amount of ＄1, 000.（因你方装船质量与所商定规格不符，因此我方提出索赔 1000 美元。）

⑤ The goods we ordered on Jun. 21, 2016 have arrived in a damaged condition.（我方于 2016 年 6 月 21 日订购的货物到货，但已受损。）

2. 求职信（Application Letter）

Dear sir,

I saw your advertisement for a (an) ________（职位）in the ________（媒体名称）of ________（日期）and I would like to apply for this position.

I would be grateful if you could send me an application form and further information about the salary and working conditions.

I look forward to hearing from you soon.

Yours faithfully,

________（签名）

◆常用语句◆

① I saw your advertisement for an export manager in the China Daily of July 8 and I would like to apply for the position.（我在七月八日 的《中国日报》上看到贵公司招聘出口部门经理的广告，于是向您申请这一职位。）

② I would be grateful if you send me an application form and further information about the salary and working conditions.（如果贵公司能惠寄求职表格并告知有关薪水和工作条件的情况，本人将不胜感激。）

③ I look forward to hearing from you in the near future.（期盼早日回复。）

Task Ⅳ

1. You are required to write a complaint letter according to the information given in Chinese below.

假想你叫王兰，上月去广州出差时，在一家商店买了一架照相机，并在广州拍了一些照片。回家后将胶卷冲洗出来，却发现什么也没照上，因此非常恼火。今天(2016 年 7 月 23 日)给商店写信投诉，并寄回相机，坚决要求退款。

Words for Reference:

出差 be on business

冲洗 develop

退款 refund

2. You are required to write an application letter according to the following instructions given in Chinese.

说明:请以王曼丽的名义写一封求职信。

王曼丽，24 岁，毕业于龙江技术学院，主修企业管理，各门课程都优良。学过速记与打字，速度各为每分钟 90 字和 70 字。请为她拟出一份给 ABC 公司的自荐信，希望能在该公司谋得总经理秘书一职。写信的日期为 2016 年 6 月 25 日。(请注意书信的格式。)

Words for reference:

技术学院 Technical College

企业管理 Business Administration

速记与打字 shorthand and typing

Unit Eight Cloud Computing

Section One Speaking

Task Ⅰ

1. Look at the following graphic and speak out each name on Cloud Computing in Chinese.

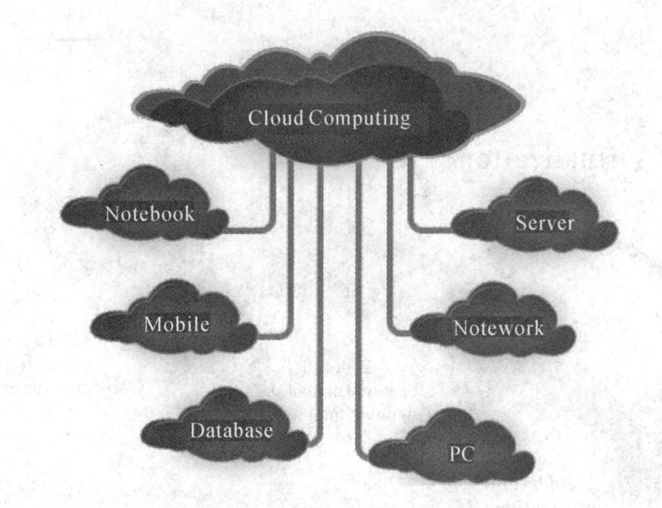

Notebook ____________ Mobile ____________ Database ____________

Server ____________ Network ____________ PC ____________

2. List the Chinese names of the following graphic.

3. Suppose you are a computer engineer. Give a presentation of each picture in English.

Picture 1

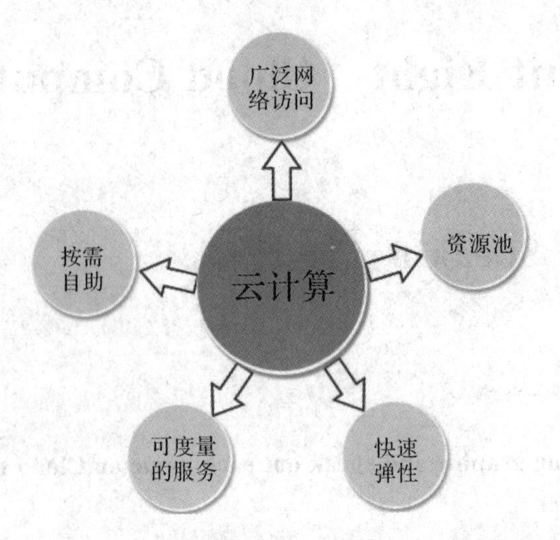

Picture 2

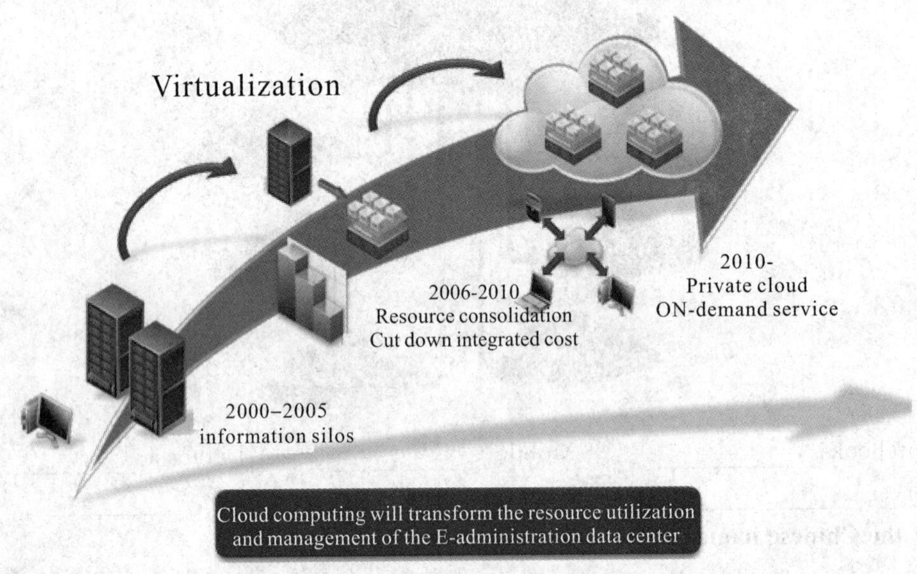

Section Two Listening

Task Ⅱ

1. Listen carefully and choose the words you hear.

(1) A. books B. boss C. bought D. boots

(2) A. cook B. cock C. caught D. cough

(3) A. foot B. fog C. fought D. food

(4) A. raw	B. rot	C. row	D. rock
(5) A. look	B. lock	C. lord	D. lawn
(6) A. took	B. top	C. talk	D. two
(7) A. show	B. shore	C. sharp	D. shock
(8) A. pill	B. pale	C. pile	D. pull
(9) A. mood	B. male	C. more	D. mall
(10) A. fill	B. fail	C. file	D. nail

2. Listen to the short passage and choose the proper words to fill in the blanks.

It's a __1__ (likely, lovely, lately) day, and it's wonderful especially after the __2__ (horrible, comfortable, terrible) storm last night. The air is fresh and the grass and trees look so green. I hope it will be like this for the __3__ (whole, so, hole) day, but the morning forecast said that we would have a __4__ (power, flower, shower) this afternoon, and the temperature would drop to __5__ (30 ℃, 13 ℃, 3 ℃). The weather changes so quickly at this time of year, so we'd better stay at home and play cards.

3. Listen to the questions and choose the best responses to what you hear.

(1) A. No, it's usually cool at this time of year.
B. The weatherman says it's going to be cloudy.
C. Sunny and warmer.
D. It seems to be clearing up. ________

(2) A. It's about 30℃.
B. It looks like snow.
C. It's quite cool here in August.
D. I doubt it. It's clear in the east and south. ________

(3) A. It says: first cloudy and windy, then rainy.
B. It was cold yesterday.
C. Yes, it will snow soon.
D. We'd better stay at home. ________

(4) A. Partly cloudy.
B. It's clearing up.
C. Yes, it's going to be fine tomorrow.
D. About 35℃, and tomorrow will be even hotter. ________

(5) A. It often snows in winter.
B. Yes, it's very cold.
C. I think it's too cold in winter.
D. I like going camping in summer. ________

(6) A. He is never late for school.
B. He has finished his report.
C. He will be better soon.
D. He's got a bad cold due to the terrible weather yesterday. ________

(7) A. It was rainy yesterday.

B. Don't bother. The weather report said it would clear up soon.

C. It will be rainy tomorrow.

D. It's quite hot here in September. ________

(8) A. Yes, it's really very cold.

B. I hope it will be sunny tomorrow.

C. Maybe 20 below.

D. No, we'd better stay indoors. ________

4. Listen to the dialogues and choose the best answer to each question you hear.

(1) A. They had lost their way. B. They were told it would rain.

C. They were caught in the rain. D. They had taken an umbrella.

(2) A. Snowy. B. Cloudy.

C. Windy. D. Rainy.

(3) A. Warm. B. Cold.

C. Hot. D. Wet.

(4) A. It has stopped snowing. B. It's going to snow.

C. It has just begun to snow. D. It's still snowing.

(5) A. It's cold. B. It's hot.

C. It's snowing. D. It's raining.

5. Listen to the dialogue and answer the following questions by filling in the blanks.

(1) What was the weather like last winter?

It was cold and there was ________.

(2) Which season is the most beautiful in a year?

________.

(3) Which month was the nicest last spring?

________.

(4) What's the weather like in July?

It is ________.

(5) In which month did we have lots of rain?

In ________.

6. Listen to the passage and fill in the blanks with words you hear. Some new words given below will be of some help to you.

skin	[skɪn]	n.	皮，皮肤
Innuits		n.	因纽特人
North Pole			北极

The Innuits

The Innuits live __1__ the North Pole. There are only two seasons there, __2__ and __3__. There are no spring or autumn there. The winter nights are __4__. You can't see the ___5___ for more than two months, even at noon. The summer days are long. For more than two months, the sun never ___6___ and there is no night.

The Innuits have ___7___ clothes. They made their clothes from the skins of ___8___. From skins they make coats, caps and shoes.

Near the North Pole trees can't grow, for it is too cold there. The Innuits have to make their ___9___ from skins, stones or snow. When they go out in a storm and can't get back home, they make houses of snow. They ___10___ these snow houses when the storm is over.

Life is hard for the Innuits, but they still like to live there.

Section Three Reading

Passage A

Cloud Computing

Cloud computing is a form of Internet-based computing that provides shared computer processing resources and data to computers and other devices on demand. It is a model for enabling ubiquitous, on-demand access to a shared pool of configurable computing resources (e. g. , computer networks, servers, storage, applications and services), which can be rapidly provisioned and released with minimal management effort. Basically, Cloud computing allows the users and enterprises with various capabilities to store and process their data in either privately owned cloud, or on a third-party server in order to make data accessing mechanisms much more easy and reliable. Data centers are that may be located far from the user — ranging in distance from across a city to across the world. Cloud computing relies on sharing of resources to achieve coherence and economy of scale, similar to a utility (like the electricity grid) over an electricity network.

Advocates claim that cloud computing allows companies to avoid up-front infrastructure costs (e. g. , purchasing servers). As well, it enables organizations to focus on their core businesses instead of spending time and money on computer infrastructure. Proponents also claim that cloud computing allows enterprises to get their applications up and running faster, with improved manageability and less maintenance, and enables information technology (IT) teams to more rapidly adjust resources to meet fluctuating and unpredictable business demand. Cloud providers typically use a "pay as you go" model. This will lead to unexpectedly high charges if administrators do not adapt to the cloud pricing model.

In 2009, the availability of high-capacity networks, low-cost computers and storage devices as well as the widespread adoption of hardware virtualization, service-oriented architecture, autonomic and utility computing led to a growth in cloud computing.

Companies can scale up as computing needs increase and then scale down again as demands decrease. In 2013, it was reported that cloud computing had become a highly demanded service or utility due to the advantages of high computing power, cheap cost of services, high performance, scalability, accessibility as well as availability. Some cloud vendors are experiencing growth rates of 50% per year. But being still in a stage of infancy, it has pitfalls that need to be addressed to make cloud computing services more reliable and user friendly.

★ ***New Words***

demand [dɪ'mɑ:nd] vt. 要求；需要；查询 vi. 请求；查问 n. [经] 需求
ubiquitous [ju:'bɪkwɪtəs] adj. 普遍存在的；无所不在的
pool [pu:l] n. 联营；撞球；共同资金 vi. 联营，合伙经营
configurable [kən'fɪgjərəbl] adj. 可配置的；结构的
rapidly ['ræpɪdlɪ] adv. 迅速地；很快地；立即
provision [prə'vɪʒ(ə)n] n. 规定；准备 vt. 供给……食物及必需品
release [rɪ'li:s] vt. 释放；发射；让与 n. 释放；发布；让与
effort ['efət] n. 努力；成就
distance ['dɪstəns] n. 距离；远方；疏远；间隔 vt. 疏远
achieve [ə'tʃi:v] vt. 取得；获得；实现；成功 vi. 达到预期的目的
coherence [kə(ʊ)'hɪər(ə)ns] n. 一致；连贯性；凝聚
economy [ɪ'kɒnəmɪ] n. 经济；节约；理财
scale [skeɪl] n. 规模；比例；刻度；天平；数值范围 vi. 衡量
utility [ju:'tɪlɪtɪ] n. 实用；公共设施；功用 adj. 实用的；通用的
grid [grɪd] n. 网格；格子，栅格；输电网
advocate ['ædvəkeɪt] vt. 提倡，主张，拥护 n. 提倡者；支持者；律师
claim [kleɪm] vi. 提出要求 vt. 要求；声称；需要；认领 n. 要求
up-front ['ʌpfrʌnt] adj. 预先的；坦率的
infrastructure ['ɪnfrəstrʌktʃə] n. 基础设施；公共建设；下部构造
organization [ˌɔ:gənaɪ'zeɪʃn] n. 组织；机构；体制；团体
proponent [prə'pəʊnənt] n. 支持者；建议者；提出认证遗嘱者
manageability [ˌmænɪdʒə'bɪləti] n. 易处理；易办；顺从
adjust [ə'dʒʌst] vt. 调整，使……适合；校准 vi. 调整
fluctuate ['flʌktʃʊeɪt] vi. 波动；涨落；动摇 vt. 使波动
unpredictable [ʌnprɪ'dɪktəbl] adj. 不可预知的；不定的；出乎意料的
unexpectedly ['ʌnik'spektidli] adv. 出乎意料地，意外地
charge [tʃɑ:dʒ] n. 费用；电荷
availability [əˌveɪlə'bɪlətɪ] n. 可用性；有效性；实用性
widespread ['wʌɪdsprɛd] adj. 普遍的，广泛的；分布广的
adoption [ə'dɒpʃ(ə)n] n. 采用；收养；接受
virtualization ['vɜ:tʃʊəˌlaɪzeiʃən] n. [计] 虚拟化

architecture [ˈɑːkɪtektʃə] n. 建筑学；建筑风格；建筑式样；架构
autonomic [ˌɔːtəˈnɒmɪk] adj. 自律的；自治的
scalability [ˌskeiləˈbiliti] n. 可扩展性；可伸缩性；可量测性
accessibility [əkˌsesəˈbɪlətɪ] n. 易接近；可亲；可以得到
vendor [ˈvendəː] n. 卖主；小贩；供应商
infancy [ˈɪnf(ə)nsɪ] n. 初期；婴儿期；幼年
pitfall [ˈpɪtfɔːl] n. 陷阱，圈套；缺陷
address [əˈdres] vt. 写姓名地址；处理
reliable [rɪˈlaɪəb(ə)l] adj. 可靠的；可信赖的 n. 可靠的人

★ ***Phrases and Expressions***

cloud computing	云计算
electricity network	电网
focus on	集中于
instead of	而不是
up and running	正常运行
pay as you go	账单到期即付
service-oriented	服务导向，服务至上

Task Ⅲ - 1

1. Fill in the blanks without referring to the passage.

Cloud computing is a form of Internet-based computing that provides shared computer ________ (1) and data to computers and other devices on demand. It is a ________(2) for enabling ubiquitous, on-demand access to a shared ________(3) of configurable computing resources which can be rapidly provisioned and released with minimal ________(4). Basically, Cloud computing allows the users and ________(5) with various capabilities to store and process their data in either privately owned cloud, or on a third-party server in order to make data accessing mechanisms much more easy and ________(6).

2. Answer the following questions according to the passage.

(1) What is cloud computing?

____

(2) What does cloud computing rely on to achieve coherence and economy of scale?

____

(3) What enables organizations to focus on their core businesses instead of spending time and money on computer infrastructure?

____

____

(4) When did cloud computing begin to grow?

____

(5) Why had cloud computing become a highly demanded service or utility?

____

3. Complete each of the following statements according to the passage.

(1) ________ is a model for enabling ubiquitous, on-demand access to a shared pool of configurable computing resources.

(2) Data centers that may be located far from the user-ranging in distance from across a city to across ________.

(3) Advocates claim that cloud computing allows companies to avoid upfront ________ (e. g., purchasing servers).

(4) Cloud providers use a ________ model, which will lead to unexpectedly high charges if administrators do not adapt to the cloud pricing model.

(5) In 2013, it was reported that cloud computing had become a highly demanded service or ________.

(6) Some cloud vendors are experiencing growth rates of ________ per year.

4. Translate the following sentences into English.

(1) 这些设备与云计算的关系可能并不那么显而易见。(cloud computing)

____

(2) 不同的技能层次如何对这样的一个模式做出回应呢?(model)

____

(3) 此时,您已经设置好基础服务器,但您的应用程序还没有设置。(server)

____

(4) 为了完成这些部分的学习,您应该拥有一台正常运行的 X 服务器。(up and running)

____

(5) 该组件为用户提供一个包含可配置文本的复选框。(configurable)

____

(6) 选择你想做的工作,不要只是为了工作而工作。(instead of)

____

(7) 昨天晚上,我签署了"分期付款"规定使它成为法律。(pay as you go)

____

(8) 我应该采用哪种虚拟化?(virtualization)

____

5. Translate the following sentences into Chinese.

(1) Cloud computing is a form of Internet—based computing that provides shared computer processing resources and data to computers and other devices on demand.

____

____

(2) Data centers that may be located far from the user-ranging in distance from across a city to across the world.

____

(3) Cloud computing relies on sharing of resources to achieve coherence and economy of scale, similar to a utility over an electricity network.

____

____

(4) Advocates claim that cloud computing allows companies to avoid up-front infrastructure costs (e. g. , purchasing servers).

____

____

(5) As well, it enables organizations to focus on their core businesses instead of spending time and money on computer infrastructure.

____

____

(6) Cloud providers typically use a "pay as you go" model. This will lead to unexpectedly high charges if administrators do not adapt to the cloud pricing model.

____

____

(7) Companies can scale up as computing needs increase and then scale down again as demands decrease.

____

(8) In 2013, it was reported that cloud computing had become a highly demanded service or utility.

____

(9) Some cloud vendors are experiencing growth rates of 50% per year, but being still in a stage of infancy.

____

Passage B

The Key Characteristics of Cloud Computing

• Agility for organizations may be improved, as cloud computing may increase users' flexibility with re-provisioning, adding or expanding technological infrastructure resources.

• Cost reductions are claimed by cloud providers. A public-cloud delivery model converts capital expenditures (e. g. , buying servers) to operational expenditure. This purportedly lowers barriers to entry, as infrastructure is typically provided by a third party and need not be purchased for one-time or infrequent intensive computing tasks. Pricing on a utility computing basis is "fine-grained", with usage-based billing options.

• Device and location independence enable users to access systems using a web browser regardless of their location or what device they use (e. g. , PC, mobile phone). As infrastructure is off-site (typically provided by a third-party) and accessed via the Internet, users can connect to it from anywhere.

• Maintenance of cloud computing applications is easier, because they do not need to be installed on each user's computer and can be accessed from different places (e. g., different work locations, while travelling, etc.).

• Multitenancy enables sharing of resources and costs across a large pool of users.

• Performance is monitored by IT experts from the service provider, and consistent and loosely coupled architectures are constructed using web services as the system interface.

• Productivity may be increased when multiple users can work on the same data simultaneously, rather than waiting for it to be saved and emailed.

• Reliability improves with the use of multiple redundant sites, which makes well-designed cloud computing suitable for business continuity and disaster recovery.

• Scalability and elasticity run via dynamic ("on-demand") provisioning of resources on a fine-grained, self-service basis in near real-time, without users having to engineer for peak loads. This gives the ability to scale up when the usage need increases or down if resources are not being used.

• Security can improve due to centralization of data and increased security-focused resources. Security is often as good as or better than other traditional systems, in part because service providers are able to devote resources to solving security issues that many customers cannot afford to tackle or which they lack the technical skills to address. However, the complexity of security is greatly increased when data is distributed over a wider area or over a greater number of devices, as well as in multi-tenant systems shared by unrelated users. In addition, user access to security audit logs may be difficult or impossible. Private cloud installations are in part motivated by users' desire to retain control over the infrastructure and avoid losing control of information security.

★ ***New Words***

key [ki:] n. (打字机等的)键；关键；钥匙 vt. 键入
characteristic [kærəktəˈrɪstɪk] adj. 典型的；特有的 n. 特征；特性
flexibility [ˌfleksɪˈbɪlɪtɪ] n. 灵活性；弹性；适应性
capital [ˈkæpɪt(ə)l] n. 首都，省会；资金，重要的
expenditure [ɪkˈspendɪtʃə; ek-] n. 支出，花费；经费，消费额
purportedly [ˈpɜ:pɒtɪdli] adv. 据称，据称地
lower [ˈləʊə] vt. 减弱，减少，降下；贬低 vi. 降低；减弱；跌落
barrier [ˈbærɪə] n. 障碍物，屏障；界线 vt. 把……关入栅栏
infrequent [ɪnˈfri:kw(ə)nt] adj. 罕见的；稀少的；珍贵的；不频发的
one-time [ˈwʌntaɪm] adj. 以前的；一次性的 adv. 一度；从前
intensive [ɪnˈtensɪv] adj. 加强的；集中的；透彻的；加强语气的
fine-grained [ˈfain-ˈgreind] adj. 细粒的 n. 细粒度

multitenancy　[ˌmʌtlɪ'tenəsɪ] n. 多租户技术
consistent　[kən'sɪst(ə)nt] adj. 始终如一的，一致的；坚持的
loosely　['luːslɪ] adv. 宽松地；放荡地；轻率地
coupled　['kʌpld] adj. 耦合的；联结的；成对的
construct　[kən'strʌkt] vt. 建造，构造；创立，搭建
interface　['ɪntəfeɪs] n. 界面；接口；接触面
productivity　[prɒdʌk'tɪvɪtɪ] n. 生产力；生产率；生产能力
multiple　['mʌltɪpl] adj. 多重的；多样的；许多的
simultaneously　[ˌsɪml'teɪnɪəslɪ] adv. 同时地
redundant　[rɪ'dʌnd(ə)nt] adj. 多余的，过剩的；冗长的，累赘的
well-designed　['weldi'zaind] adj. 精心设计的；设计巧妙的
suitable　['suːtəb(ə)l] adj. 适当的；相配的
continuity　[ˌkɒntɪ'njuːɪtɪ] n. 连续性；一连串
scalability　[ˌskeilə'biliti] n. 可扩展性；可伸缩性；可量测性
elasticity　[elæ'stɪsɪtɪ] n. 弹性；弹力；灵活性
dynamic　[daɪ'næmɪk] adj. 动态的；动力的；有活力的　n. 动力
provisioning　[prə'vɪʒən] n. 准备金提取　v. 供应补给品
real-time　[ˌrɪəl 'taɪm] adj. 实时的；接到指示立即执行的
centralization　[ˌsentrəlɪ'zeʃən] n. 集中化；中央集权管理
tackle　['tæk(ə)l] n. 装备；用具　vt. 处理；扭倒
distribute　[dɪ'strɪbjuːt] vt. 分配；散布；分开
retain　[rɪ'teɪn] vt. 保持；雇；记住

★ ***Phrases and Expressions***

capital expenditure	基本建设费用
look into	调查，观察
web browser	网络浏览器
regardless of	不管，不顾
rather than	而不是
scale up	按比例
in part	在某种程度上

Task Ⅲ－2

1. Match the following English phrases in Column A with their Chinese equivalents in Column B.

A	B
(1) electricity network	a. 服务导向
(2) up and running	b. 云计算
(3) cloud computing	c. 电网

(4) pay as you go　　d. 正常运行
(5) business demand　　e. 账单到期即付
(6) service-oriented　　f. 商业需求
(7) web browser　　g. 基础设施资源
(8) capital expenditure　　h. 按比例
(9) scale up　　i. 基本建设费用
(10) infrastructure resource　　j. 网络浏览器

2. Decide whether the following statements are T(true) or F (false) according to the passage.

(1) Cloud computing may increase users' flexibility with re-provisioning, adding or expanding technological infrastructure resources. (　　)

(2) Cost increase are claimed by cloud providers. (　　)

(3) In cloud computing, device and location independence enable users to access systems using a web browser regardless of their location or what device they use. (　　)

(4) Maintenance of cloud computing applications is not easy, because they need to be installed on each user's computer. (　　)

(5) In cloud computing, multitenancy doesn't enable sharing of resources and costs across a large pool of users. (　　)

(6) Performance is monitored by IT experts from the service provider in cloud computing. (　　)

(7) Productivity may be increased when multiple users can work on the same data simultaneously, rather than waiting for it to be saved and emailed in cloud computing. (　　)

(8) Cloud computing isn't suitable for business continuity and disaster recovery. (　　)

(9) Scalability gives the ability to scale up when the usage need increases or down if resources are not being used. (　　)

(10) Security can improve because of centralization of data, increased security-focused resources. (　　)

3. Translate the following passage into Chinese.

Cost reductions are claimed by cloud providers. A public-cloud delivery model converts capital expenditures (e. g. , buying servers) to operational expenditure. This purportedly lowers barriers to entry, as infrastructure is typically provided by a third party and need not be purchased for one-time or infrequent intensive computing tasks. Pricing on a utility computing basis is "fine-grained", with usage-based billing options.

Read the following passage and choose the best answer to each question.

Deployment Models

Private cloud

Private cloud is cloud infrastructure operated solely for a single organization, whether managed internally or by a third-party, and hosted either internally or externally. Undertaking a private cloud project requires significant engagement to virtualize the business environment, and requires the organization to reevaluate decisions about existing resources. It can improve business, but every step in the project raises security issues that must be addressed to prevent serious vulnerabilities. Self-run data centers are generally capital intensive. They have a significant physical footprint, requiring allocations of space, hardware, and environmental controls. These assets have to be refreshed periodically, resulting in additional capital expenditures. They have attracted criticism because users "still have to buy, build, and manage them" and thus do not benefit from less hands-on management, essentially the economic model that makes cloud computing such an intriguing concept.

Public cloud

A cloud is called a "public cloud" when the services are rendered over a network that is open for public use. Public cloud services may be free. Technically there may be little or no difference between public and private cloud architecture, however, security consideration may be substantially different for services (applications, storage, and other resources) that are made available by a service provider for a public audience and when communication is effected over a non-trusted network. Generally, public cloud service providers like Amazon Web Services (AWS), Microsoft and Google own and operate the infrastructure at their data center and access is generally via the Internet. AWS and Microsoft also offer direct connect services called "AWS Direct Connect" and "Azure Express Route" respectively, such connections require customers to purchase or lease a private connection to a peering point offered by the cloud provider.

Hybrid cloud

Hybrid cloud is a composition of two or more clouds (private, community or public) that remain distinct entities but are bound together, offering the benefits of multiple deployment models. Hybrid cloud can also mean the ability to connect collocation, managed and/or dedicated services with cloud resources. Gartner, Inc. defines a hybrid cloud service as a cloud computing service that is composed of some combination of private, public and community cloud services, from different service providers. A hybrid cloud service crosses isolation and provider boundaries so that it can't be simply put in one category of private, public, or community cloud service. It allows one to extend either the capacity or the capability of a cloud service, by aggregation, integration or customization with another cloud service.

1. Which is not true about private cloud in the following statement?

   A. Private cloud is cloud infrastructure operated solely for a single organization.

B. Private cloud is whether managed internally or by a third-party.

C. Undertaking a private cloud project doesn't require significant engagement to virtualize the business environment.

D. Private cloud is hosted either internally or externally.

2. Which is not true about self-run data centers in the following statement?

A. They have a significant physical footprint.

B. They are generally capital intensive.

C. They require allocations of space, hardware, and environmental controls.

D. They don't have attracted criticism.

3. What is public cloud according to the passage?

A. is a composition of two or more clouds that remain distinct entities but are bound together, offering the benefits of multiple deployment models.

B. Public cloud is cloud infrastructure operated solely for a single organization.

C. A cloud is called a "public cloud" when the services are rendered over a network that is open for public use.

D. Technically there are a lot of difference between public and private cloud architecture.

4. What may be substantially different between public and private cloud architecture?

A. Technique may be substantially different for services between public and private cloud architecture.

B. Security consideration may be substantially different for services between public and private cloud architecture.

C. Maintenance may be substantially different for services between public and private cloud architecture.

D. Usage may be substantially different for services between public and private cloud architecture.

5. Hybrid cloud is ________ that remain distinct entities but are bound together, offering the benefits of multiple deployment models. ?

A. a composition of two or more clouds

B. a composition of two clouds

C. a composition of private and community cloud

D. a composition of community and public cloud

Section Four　Writing

通知和启示(Notice & Announcement)

1. 通知(Notice)

通知通常是上级对下级、组织和团体对其成员部署工作或传达事项等所使用的一种公告性文件。发通知的一方和被通知的一方一般以第三人称出现。如果正文前用了称呼，通知的对象用第二人称。

Notice

Aug. 10th，2016

Your attention，please. A visit has been arranged to the Great Wall for October 15. Foreign students who wish to go please assemble at the Students' Club. The coach is to leave at 8 a. m.

Foreign Affairs Office

通　知

请注意：定于十月十五日参观长城。想去的留学生请在学生俱乐部集合。校车上午八点出发。

外事处

2016 年 8 月 10 日

2. 启示（Announcement）

启示是一种公告性应用文字。机关、团体、单位或个人如有事情向公众说明，或有什么请求，可以把要说的内容简要地写成启事。启事不用称谓，但有的启事需写清地址。

Announcement

China Daily welcomes letters from readers on all local，national and international subjects. Preference will be given to short letters. Letters should be mailed to：Letter-to-the-Editor，China Daily，15 Huixing Dongjie，Beijing，100029

启　示

《中国日报》欢迎读者就各个地方性的、国内国际的问题来信，最好是言简意赅的短信。来信请寄：北京 100029，汇兴东街 15 号《中国日报》"读者来信"编辑组收。

Task Ⅳ

1. The following is a contribution（征稿）for an English Corner in Chinese local newspaper. Put both the introduction and the contributions into English. The title and the first sentence have been given to you.

征稿启事

"英语角"在大学生专刊上刊发后，受到广大英语读者的广泛关注和欢迎。为刊发更多更好的优秀英文作品，我们特向社会各界征集稿件。我们欢迎大学生们、广大英语学习者、外国留学生及外国朋友们踊跃投稿。征稿内容为：

（1）一则简短的故事；

（2）发生在你身边的趣事。

文章字数不要超过 500 字。

来稿请寄：北海市新民大街 10 号北海晚报王平收

邮编：230034

电子邮件：wangping@163.com

Words for Reference:

各界 all walks of life

2. You are required to write an Announcement according to the following information in Chinese.

(1) 事由：欢迎美国学生来我校参观；

(2) 时间：2016 年 6 月 22 日上午 9:00 至 13:00；

(3) 人数：约 35 人；

(4) 具体安排：

时间	地点	事项
8:40	校门口	集合；欢迎客人
9:00—10:30	接待室	开欢迎会；互赠礼物
10:30—11:30	校园、图书馆、实验室和语言实验室	参观
11:30	学生食堂	午餐
13:00	校门口	送别

Words for Reference:

接待室 reception room

欢迎会 welcome party

食堂 student canteen

3. The following contains the main information of a notice of the Public Relations Department of a joint-venture(合资企业). You are required to write an English notice based on the following points.

说明：根据下列信息以公关部的名义给所有员工写一份公告，邀请他们为公司庆祝活动献计献策。

(1) 历史与现状：成立 15 年，在规模和效益方面现处于同行业五强之一；

(2) 庆祝活动：举行一系列活动，庆祝取得的成就；

(3) 欢迎献计献策：被采用者有奖，所提建议送往本部门办公室。

Words for Reference:

规模 scope

经济效益 economic benefits　　同行业 the same industry

同仁 colleague　　献计献策 make proposals

Appendix A Listening Scripts

Unit One Computer Hardware

1. Listen carefully and choose the words you hear.

(1) dad (2) food (3) please (4) sleep (5) chemical
(6) form (7) dig (8) guess (9) board (10) kid

2. Listen to the short passage and choose the proper words to fill in the blanks.

American and British people use different greetings. In the USA, the commonest greeting is "Hi ". In Britain it is "Hello!" or "How are you? ". "Hi! " is spreading into British, too. When they are introduced to someone, the Americans say, "Glad to know you ". The British say, "How do you do? " or "Pleased to meet you. " When Americans say "Goodbye ", they nearly always add, "Have a good day. " or "Have a good trip. " etc. to friends and strangers alike. The British are already beginning to use "Have a good day. "

3. Listen to the dialogues and choose the best responses to what you hear.

(1) M: Let me introduce myself. I'm Bob.

(2) M: Nice to meet you. I'm Jack from Australia.

(3) W: Mr. Li, I'd love to introduce myself. My name is Lucy Evans, a secretary of company.

(4) W: I'd like to introduce my friend Tom Green to you. Tom, this is Mr. Lin.
M: How do you do you, Mr. Lin?

(5) W: This is my brother Sun Hao.

(6) M: Linda, this is my sister Jane.
W: I'm very glad to meet you, Jane.

(7) M. Oh, Helen, I'd like you to meet my classmate Linda.
W: Hi. Linda.

(8) M: Hello, I'm Bill. I'm your neighbor.
W: I'm Jenny. Nice to meet you.

4. Listen to the dialogues and choose the best answer to each question you hear.

(1) W: Hi, Alice. Is Mary twelve years old?
M: No, she's thirteen and Sue is twelve years old.
Q: Who is younger?

(2) M: Let's go out to play.

W: But it's snowing now.

Q: What season is it?

(3) W: What do you usually do at the weekend, Ben?

M: I usually play football. I sometimes go to the cinema.

Q: What does Ben usually do at the weekend?

(4) W: We're going to visit the Space Museum.

M: By underground or by bus?

W: By bus. It's cheaper.

Q: How are they going to the Space Museum?

(5) W: When is the train leaving for Tianjin?

M: At 3 p. m. What time is it now?

W: It's about a quarter to three. Let's hurry.

Q: What time is the train leaving?

5. Listen to the dialogue and answer the following questions by filling in the blanks.

W : Aha, there he is.

M1: Jenny, How nice to see you again. How are you?

W : Fine, thank you. And you?

M1: Fine, thanks. May I introduce you to our manager, Liu Wei, who is here to meet you.

W : How do you do, Mr. Liu.

M2: How do you do, Jenny? Welcome to China.

W : Thank you very much. I've been looking forward to this trip. It's very kind of you to invite me.

M2: It's my pleasure. Did you have a good trip?

W : Not bad on the whole, but you know it's a long way to China. We were flying for about nine hours.

M2: Then you must be tired.

W : Well, not really. I manage to get some sleep.

6. Listen to the passage and fill in the blanks with words you hear. Some new words given below will be of some help to you.

My Experiences in College

Like many other freshman, I showed great interest in college life when I arrived at the college. I saw new faces, made new friends, went to new places, and had new classes. Everything seemed new and interesting to me.

One week later, we began to have classes. On weekdays, I had many classes, such as mathematics, business and English. On weekends, I would climb the mountains or go to the beach with my classmates by bicycle. Sometimes we had a picnic and I usually had a good time there. When we were about to graduate, we held a lot of parties on campus to celebrate our graduation and show our friendship to each other.

Now five years have passed, but college life is still fresh in my mind and I still cherish those good old days at college. I miss my classmates very much, and I dream that all of our classmate can meet again at college some day.

Unit Two Computer Software

1. Listen carefully and choose the words you hear.

(1) blow (2) get (3) clean (4) vase (5) fat

(6) forget (7) kind (8) dot (9) home (10) taught

2. Listen to the short passage and choose the proper words to fill in the blanks.

Linda is my best friend. She is 15 years old. She is a pretty girl with a round face and two big black eyes. She always has a smile on her face. She is taller than I.

Every morning we go to school together. She studies quite well and she's a top student in our class. She is modest in her behavior. When I have difficulty in English, I always ask her for help. We are both interested in music. At weekends, we join the same hobby group and play the violin together. We like each other.

3. Listen to the dialogues and choose the best responses to what you hear.

(1) M: How wonderful the plane show is!

W: Yes, but do you know how fast the first plane flew?

(2) M: I'm going to have a picnic today. What's the weather like?

(3) M: I can't find my book.

W: Don't worry. Let me help you. Oh, it's here, under the desk.

M: Thank you for your help.

(4) M: I've been to Beijing a few days before.

W: Really? It's a good place! Have you been to the Great Wall?

(5) M: I'd like to eat a hamburger. What about you?

4. Listen to the dialogues and choose the best answer to each question you hear.

(1) M: Does your father still stay in America, Linda?

W: He works there as a Chinese teacher in a middle school.

Question: What's Linda's father?

(2) W: How did you like your trip in Hong Kong?

M: Oh, I like it very much. The food is good and the people are friendly. There are many interesting places. But the weather is very hot. I didn't like it.

Question: What did the man think of his trip?

(3) M: what have you been doing these past three years?

W: That's really been that long? How time flies?

Question: What's their relationship?

(4) M: Do you come to this place often?

W: I used to. But recently I've been working on a lot of overtime, so I don't have much time to enjoy myself.

Question: Where did the conversation take place?

(5) M: Would you like to be a teacher like your mother, Jane?

W: No. I want to be a doctor.

Question: What does Jane's mother do?

5. Listen to the dialogue and answer the following questions by filling in the blanks.

M: Li Hua, is that you?

W: Yes. You are Wang Hai, aren't you?

M: Yes, that's right. You have changed a lot!

W: Really? How?

M: You used to be short. Now you're tall.

W: Yes, I am.

M: You used to have short hair. Now you have long hair.

W: Yes, that's right.

M: And you used to wear glasses.

W: Yes, I still do. But today I have contact lenses.

M: You used to play the piano, didn't you?

W: No, I played the guitar. My friend Jerry played the piano.

M: Oh, yes. I forgot.

6. Listen to the passage and fill in the blanks with words you hear. Some new words given below will be of some help to you.

Old Couple at McDonald's

A little old man and his wife walked slowly into McDonald's on cold winter evening. They took a table near the back wall, and then the little old man walked to the cash register to order. After a while he got the food back and they began to open it.

There was one hamburger, some French fries and one drink. The little old man carefully cut the hamburger in half and divided the French fries in two piles. Then he neatly put the half of the food in front of his wife. He took a sip of the drink and his wife took a sip. "How poor the old people are!" The people around them thought. As the man began to eat his hamburger and his French fries, his wife sat there watching him and took turns to drink. A young man came over and offered to buy another meal for them. But they refused politely and said that they got used to sharing everything.

Then a young lady asked a question to the little old lady. "Madam, why aren't you eating? You said that you share everything. Then what are you waiting for?" She answered, "The teeth."

Unit Three Computer Network

1. Listen carefully and choose the words you hear.

(1) cloud (2) pass (3) weather (4) shoot (5) right

(6) fork (7) whose (8) methods (9) whole (10) sink

2. Listen to the short passage and choose the proper words to fill in the blanks.

GSM (Global System for Mobile Communications) is the most popular standard for mobile phones in the world. Its promoter, the GSM Association, estimates that 80% of the global mobile market uses the standard. GSM is used by over 3 billion people across more than 212 countries and territories. Its ubiquity makes international roaming very common between mobile phone operators, enabling subscribers to use their phones in many parts of the world. GSM differs from its predecessors in that both signaling and speech channels are digital and thus is considered a second generation (2G) mobile phone system. This has also meant that digital communication was easy to build into the system.

3. Listen to the dialogues and choose the best responses to what you hear.

(1) A. When are you flying back?

(2) Operator: 114 Bell, may I help you?

(3) A: What's the weather today?

(4) Man: What a beautiful dress you have on!

(5) A: So, we'll have the Christmas off. Have a nice holiday.

(6) A: Excuse me. Can you tell me the way to the Holiday Inn?

(7) A: Hi, how are you doing?

B: I'm doing well. How about you?

(8) A. Thank you very much for inviting us to such a wonderful party.

4. Listen to the dialogues and choose the best answer to each question you hear.

(1) W: I think a teacher should be a friendly person.

M: Yes, I agree.

Q: What does the woman think a teacher should be?

(2) M: Have you heard the news?

W: No.

M: There's been a terrible air crash.

W: Which company?

M: Air France.

Q: What is the news about?

(3) W: Excuse me, how do you say the word, c-o-m-m-u-n-i-c-a-t-e?

M: Communicate.

W: I see, thank you.

Q: What are they talking about?

(4) M: Would you like something to drink?

W: Just some juice, please.

M: How about some coffee? I've made wonderful coffee.

W: No, thanks.

Q: What does the woman want?

(5) M: What are you doing in New York?

W: I'm writing stories for a magazine.

M: I see.

Q: What is she doing there?

5. Listen to the dialogue and answer the following questions by filling in the blanks.

Reporter : Is it true that you don't swim at all now?

Alice White : I'm afraid so. I'm too old.

Reporter : But you are only 20.

Alice White : That's too old for a swimmer. If I swam in an international competition now, I wouldn't win. So I'd rather not swim at all.

Reporter : But don't you enjoy swimming?

Alice White : I used to, when I was small. But if you enter for big competitions you have to work very hard. I used to get up at 6 a. m. to go to the pool. I had to train before school, after school and at weekends. I swam 35 miles every week.

Reporter : But you were famous at 15. And look at all those cups.

Alice White : It's true that I have some wonderful memories. I enjoyed visiting other countries, and the Olympics were very exciting. But I missed more important things. While other girls were growing up, I was swimming. What can I do now?

6. Listen to the passage and fill in the blanks with words you hear. Some new words given below will be of some help to you.

How Babies Communicate ?

Babies are born with the ability to cry, which is how they communicate for a while. Your baby's cries generally tell you that something is wrong: an empty belly, a wet bottom, cold feet, being tired, or a need to be held and cuddled, etc.

Soon you'll be able to recognize which need your baby is expressing and respond accordingly. In fact, sometimes what a baby needs can be identified by the type of cry-for example, the "I'm hungry" cry may be short and low-pitched, while "I'm upset" may sound choppy.

Your baby may also cry when overwhelmed by all of the sights and sounds of the world, or for no apparent reason at all. Don't be too upset when your baby cries and you aren't able to console him or her immediately: crying is one way babies shout out stimuli when they're overloaded.

Unit Four Multimedia

1. Listen carefully and choose the words you hear.

(1) great (2) low (3) played (4) why (5) orange
(6) worker (7) Sunday (8) her (9) how (10) heard

2. Listen to the short passage and choose the proper words to fill in the blanks.

For many people, a birthday is one of the most important days of the year. And it is time for celebrations with families and friends. On this day, they have made much delicious food, such as ice cream, cake and sandwiches. They can also receive many gifts. Opening the gifts is the nicest part of the party. They are singing, dancing and having a good time. Before they eat the cake, the birthday child will blow out candles and make a wish.

3. Listen to the dialogues and choose the best responses to what you hear.

(1) M: Hi, Peter, this is my new friend, Tom.
(2) W: How are you?
M: I'm fine, thank you, and you?
(3) M: I want to speak to Wu Dong, please.
(4) M: What's in the bag?
(5) M: How often do you watch TV?
(6) W: Thank you for your help.
(7) W: What's wrong with you?
(8) M: Would you like to come?

4. Listen to the dialogues and choose the best answer to each question you hear.

(1) W: We do need another bookshelf in this room, but the problem is the space for it.
M: How about moving the old dinner table to the kitchen?
Q: What does the man suggest the woman to do?
(2) M: If I were you, I'd like a plane instead of bus. It will take you a whole day to get there.
W: But flying makes me so nervous.
Q: What does the woman prefer to do?
(3) W: I want to look at some hats.
M: OK, here you are. We have all kinds and sizes.
Q: Where are the man and woman talking?
(4) M: If there is anything wrong, please call me back.

W: OK, sir. I will look after your baby well.

Q: What does the woman do?

(5) W: It's already 7: 00. You should have been here half an hour ago.

M: I am sorry. I met a car accident.

Q: When should the man arrive?

5. Listen to the dialogue and answer the following questions by filling in the blanks.

Waiter: Here are the menus, please.

Mr Wang: Thanks. Let me have a look.

Waiter: Ready to order?

Mr Wang: Yes, I think I would like a steak with a green salad.

Waiter: Fine. Anything to drink? Bear or wine?

Mr. Wang: I like wine better.

Waiter: OK. I will be back soon.

6. Listen to the passage and fill in the blanks with words you hear. Some new words given below will be of some help to you.

Different eating habits

Different countries have different eating habits. And I'd like to say something about it.

Firstly, Chinese people eat a lot of freshly-cooked rice, noodles, vegetables, meat, and soup. And the food is often cooked in many ways. Western people prefer bread, milk and salad. They also eat a lot of fast food like hamburgers and potato chips. Secondly, Westerners are used to eating meals with a knife and a fork, and Chinese use chopsticks. Thirdly, Westerners think it's rude to make noise during the meal, but it is common for Chinese to talk with friends when having dinner.

Unit Five E-commerce

1. Listen carefully and choose the words you hear.

(1) next (2) slum (3) five (4) fold (5) horse

(6) neither (7) throw (8) bees (9) pretty (10) silver

2. Listen to the short passage and choose the proper words to fill in the blanks.

There's No Better Way to Get in the Game!

High Definition Television is the best way to watch sports. And now you can get virtual front row seats thanks to Haier and the NBA. Haier, one of the world's leading manufacturers of HDTVs, is now the "Official HDTV of the NBA" as well as an "Official Marketing Partner" of the NBA.

Haier is always ready to jump on the course. Whether playing the game for thousands of fans or creating the technology to broadcast it to millions, both Haier and the NBA agree that high definition is ideal for NBA games. So visit this site often, get in on the action, and watch for your chance to win!

3. Listen to the dialogues and choose the best responses to what you hear.

(1) It's just the one I want. How much is it?

(2) Can I help you?

(3) I'm looking for a dress for my mother.

(4) Have you any apples?

(5) I'd like to try it on. Where is the fitting room?

(6) May I use my credit card?

(7) Do come again, please.

(8) Is there anything you would like?

4. Listen to the dialogues and choose the best answer to each question you hear.

(1) A: If this dress doesn't fit, may I bring it back later?

B: Sorry, We don't take returns on sale items.

Q: Can she refund it?

(2) A: Would you like to pay it in cash or by credit card? We give a ten percent discount for cash payment.

B: Then I'd like to pay in cash.

Q: How is he going to pay?

(3) A: Is there anything I can do for you?

B: I'm looking for a blue, leather handbag.

Q: What does she want to buy?

(4) A: Do you have any English newspapers?

B: Yes, just over there.

Q: What does B mean?

(5) A: Is somebody taking care of you?

B: No. I'd like a long-sleeved shirt in yellow, medium.

A: I think we're out of your size.

Q: What size does she want?

5. Listen to the dialogue and answer the following questions by filling in the blanks.

A: This one fits you very much.

B: I think so. How much?

A: 300 yuan.

B: Oh! It's too expensive. Can it be discounted?

A: 10% discount, 270 yuan.

B: Very good! Please pack it for me!

A: Please pay it in the paying office.

B: Can I pay it with credit card?

A: Sorry, we only accept cash here.

B: OK, may I have the receipt?

A: Of course, this is your receipt. Welcome next time!

6. Listen to the passage and fill in the blanks with words you hear. Some new words given below will be of some help to you.

Supermarket—Convenient Shopping Area

Nowadays, people's living standard has been enormously improved, and with the improvement of people's living standard, supermarkets are springing up like mushrooms. Every day, especially at weekends, crowds of shoppers flock towards supermarkets and smilingly come out with rich harvests. The enthusiasm of the shoppers is induced by the superiorities of the supermarket.

Firstly, it is very convenient to do shopping in the supermarket. What you need to do is walk your handcart along the aisles and take anything you need from the shelves to fill your cart.

Secondly, the market provides a bumper variety of goods. Food , clothes, daily articles, drinks, books, home electric appliances, all are within your arm's reach.

Thirdly, the environment is clean and comfortable with all-year-round air-condition.

Fourthly, payment is made easily, either in cash or with a card. And you don't have to wait long.

Naturally, people like shopping in supermarkets even they sometimes buy a few unnecessary things. But the next time they do shopping, they completely forget it.

Unit Six Computer Security

1. Listen carefully and choose the words you hear.

(1) better (2) bear (3) cuff (4) ship (5) cheek

(6) vest (7) vat (8) bad (9) seal (10) cot

2. Listen to the short passage and choose the proper words to fill in the blanks.

Not all people like to work but everyone likes to play. All over the world men and women, boys and girls enjoy sports. Since long, adults and children have called their friends together to spend hours, even days playing games.

Sports help people to live happily. They help to keep people healthy and feeling good. When they are playing games, people move a lot. This is good for their health. Having fun with their friends makes them happy.

Many people enjoy sports by watching others play. In small towns, crowds meet to watch the bicycle races or the soccer games. In the big cities, thousands buy tickets to see ice-skating shows or baseball games.

Is it hot where you live? Then swimming is probably your favorite sport. Boys and girls in China love to swim, while aged people like various kinds of Tai Ji and Qi Gong. The climate is good for all the seasonal sports in the country.

3. Listen to the dialogues and choose the best responses to what you hear.

(1) Do you play football as well as your father?

(2) What do you usually do when you are free?

(3) Are you a good swimmer?

(4) Do you like your PE teacher?

(5) How much time do you spend on sports each day?

(6) It's a lovely day. Why don't we go for a walk?

(7) Shall we go and watch the volleyball game tonight?

(8) Which is your favorite sport?

4. Listen to the dialogues and choose the best answer to each question you hear.

(1) W: You must have had a great time watching the football game last night.

M: Not really.

Q: How does the man feel?

(2) M: I used to take a walk every day after supper.

W: What a pity you don't do it now!

Q: What does the woman mean?

(3) M: You should do morning exercises every day.

W: I know, but I hate getting up early.

Q: What does the woman mean?

(4) M: Did the game start at 2: 00 this afternoon?

W: That's right. And when you arrived at 4: 30, it was almost over.

Q: How long did the game last?

(5) W: Jim is a good ping-pong player, and now he's beginning to practice tennis.

M: He has also practiced volleyball for three months.

Q: Which ball game does Jack play best.

5. Listen to the dialogue and answer the following questions by filling in the blanks.

W: It's so cold today.

M: Yeah, it is really.

W: Shall we still go skating? Why don't we just stay at home?

M: Well, I don't think so. We need the exercise.

W: You are right there, but we could do some exercise at home.

M: Come on, you won't feel cold when you start skating.

W: All right, then.

M: Let's go.

6. Listen to the passage and fill in the blanks with words you hear. Some new words given below will be of some help to you.

I do like sports!

I do like sports during my daily life. Talking about what kind of sport that I like most, I tell you that, it's swimming. The reasons I choose swimming are as follows:

First of all, it is really enjoyable, you know, in summer. When the sun is shining outside, and the air is even too hot to breathe, swimming would be the best choice. Once you jump into the water, the coolness will drive all the anxiety and sticky feelings away. Second, I guess swimming is one of the best sports to improve one's health. It enables your lungs to breath more air, and helps the body to build up more muscles. Last but not least, swimming is suitable for girls, since it's a whole body sport. Through the whole body's exercise, girls can gain a well-shaped figure easily.

I like sports. I like swimming.

Unit Seven Big Data

1. Listen carefully and choose the numbers you hear.

(1) 50
(2) 140
(3) 70%
(4) 200, 000, 000
(5) 13
(6) 78.21
(7) 380
(8) 70
(9) 16%
(10) 5, 500, 000, 000

2. Listen to the short passage and choose the proper words to fill in the blanks.

The Industrial Revolution, in the 18th and 19th centuries, brought a new kind of advertising. Large factories took the place of small work shops and goods were produced in large quantities. Manufactures used the newly built railroads to distribute their products over wide areas. They had to find many thousands of customers in order to stay in business. They could not simply tell the people where shoes or cloth or tea could be bought—they had to learn how to make people want to buy a specific product. Thus modern advertising was born.

3. Listen to the questions and choose the best responses to what you hear.

(1) Q: What does Mr. Brown do for a living?
(2) Q: Bill, may I use you dictionary?
(3) Q: Is there anything wrong with the car?
(4) Q: Hi, Nancy, what do you think of the movie?
(5) Q: Why did you come by taxi instead of driving your own car?
(6) Q: Excuse me, how can I get to the post office?

(7) Q: What does Mary look like?

(8) Q: How can I get to the town center?

4. Listen to the dialogues and choose the best answer to each question you hear.

(1) M: Mary, did you see my tape?

W: I put it on the table.

Q: What is the boy looking for?

(2) W: Mr. Peterson, how did the party go last night?

M: We had a wonderful time.

Q: What does the man think of the party?

(3) M: Here we are, madam. This is Beijing Hotel.

W: Thank you. How much should I pay you?

Q: What is the probable relationship between the two speakers?

(4) M: Will Dr. White's lecture begin at 1: 40 or 2 o'clock?

W: It would begin at 1: 50 and finish in 2 hours.

Q: When will the lecture begin?

(5) M: Did you hear that Mike had his arm broken?

W: Yes. He was injured in a car accident.

Q: What happened to Mike?

5. Listen to the dialogue and choose the best answer to each question you hear.

John Smith is in the sitting-room. His wife, Mary, is in the kitchen. She is calling him.

Mrs Smith: Is the baby with you, John? He's not in the kitchen.

Mr. Smith: He isn't here, Mary. Perhaps he's upstairs.

Mrs Smith: Please go and see. He's very quiet.

Mr. Smith: All right. I'll go and see. (He goes upstairs.) Mary, he's not in his room.

Mrs Smith: Is he in our room?

Mr. Smith: No, he's not here either.

Mrs Smith: Good heavens! Where's he then?

Mr. Smith: Oh, he is here. In the washroom!

Mrs Smith: In the washroom? What's he doing here?

Mr. Smith: He's cleaning his shoes with your tooth-brush!

1. Where is Mary?
2. What did Mary ask John to do?
3. What did John answer?
4. Where is the baby actually?
5. What is the baby doing?

6. Listen to the passage and fill in the blanks with words you hear. Some new words given below will be of some help to you.

Advertising

Advertising is part of our daily lives. To realize this fact, you have only to leaf through a magazine or newspaper or count the radio or television commercials that you hear in one evening. Most people see and hear several hundred advertising messages every day. And people respond to the many devices that advertisers use to gain their attention.

Advertising is a big business—and to many people, a fascinating business, filled with attraction and excitement. It is part literature, part art, and part show business.

Advertising is the difficult business of bringing information to great number of people. The purpose of an advertisement is to make people respond—to make them react to an idea, such as helping prevent forest fires, or to make them want to buy a certain product or service.

Unit Eight Cloud Computing

1. Listen carefully and choose the words you hear.

(1) boss (2) cough (3) fought (4) row (5) lock

(6) took (7) shore (8) pile (9) mood (10) nail

2. Listen to the short passage and choose the proper words to fill in the blanks.

It's a lovely day, and it's wonderful especially after the terrible storm last night. The air is fresh and the grass and trees look so green. I hope it will be like this for the whole day, but the morning forecast said that we would have a shower this afternoon, and the temperature would drop to 13℃. The weather changes so quickly at this time of year, so we'd better stay at home and play cards.

3. Listen to the questions and choose the best responses to what you hear.

(1) Q: Is the weather always like this?

(2) Q: Do you think it will rain?

(3) Q: What's the weather forecast for today?

(4) Q: What's the temperature today?

(5) Q: How do you like our weather?

(6) Q: Why didn't Paul come to school today ?

(7) Q: I left my raincoat in my room. Wait while I go back to get it.

(8) Q: How many degrees below is it?

4. Listen to the dialogues and choose the best answer to each question you hear.

(1) W: I wish we had taken an umbrella.

M: That's my fault. I thought it wouldn't rain today.

Q: What happened to the two speakers?

(2) M: The streets are covered with snow.

W: That's true. It has been snowing for a whole day.

Q: What's the weather like?

(3) M: It's very cold this morning.

W: You're right. It's much colder than yesterday.

Q: What's the weather like this morning?

(4) M: Is it still snowing outside?

W: Yes, it is. We'd better wait until it stops.

Q: What's the weather like now?

(5) M: What a terrible day, isn't it?

W: Yeah, we haven't had such cold weather for a long time.

Q: What do we know about the weather?

5. Listen to the dialogue and answer the following questions by filling in the blanks.

A: Was it cold last winter?

B: Yes.

A: Did you have much snow?

B: Yes. Of course we didn't have as much snow as you had in Geneva.

A: What about the other seasons?

B: Autumn is the most beautiful season. The days are clear and dry. The nicest months are September and October, but we had a lot of rain in November.

A: How about last spring?

B: It wasn't very nice. May was nicer than March and April. March was cold and April was wet.

A: Is it always hot in summer?

B: It wasn't hot last year. But the summer has been marvelous this year.

A: Is August as hot as July?

B: No, it's cooler and drier.

6. Listen to the passage and fill in the blanks with words you hear. Some new words given below will be of some help to you.

The Innuits

The Innuits live near the North Pole. There are only two seasons there, winter and summer. There are no spring or autumn there. The winter nights are long. You can't see the sun for more than two months, even at noon. The summer days are long. For more than two months, the sun never goes down and there is no night.

The Innuits have warm clothes. They made their clothes from the skins of animals. From skins they make coats, caps and shoes.

Near the North Pole trees can't grow, for it is too cold there. The Innuits have to make their houses from skins, stones or snow. When they go out in a storm and can't get back home, they make houses of snow. They leave these snow houses when the storm is over. Life is hard for the Innuits, but they still like to live there.

Appendix B Translation to the Passages

Unit One Computer Hardware

Passage A 计算机硬件(Ⅰ)

计算机硬件是组成计算机系统的物理元件的集合。计算机硬件即计算机的物理部件或元件，诸如显示屏、键盘、存储设备、显卡、声卡、主板等等，所有这些都是有形物体。对比之下，软件则是能够储存并通过硬件运行的指令。

计算机硬件由软件指导执行命令或指令。软件和硬件的组合形成了一个可用的计算机信息处理系统。

个人计算机

个人计算机，也称PC机，由于其多功能和相对低价位成为最常用的一种计算机类型。笔记本电脑通常与之相似，尽管笔记本电脑能耗较低且减小了元件尺寸，但也因此功效较低。

电脑机箱

电脑机箱是一个塑料或金属的、包裹着元器件的附件。台式机机箱通常较小适合置于桌下。然而，近年来更加紧缩的设计已变得司空见惯，例如来自苹果厂家的被命名为苹果一体机的一体化风格设计(即是此类型)。一个机箱可大可小，但是设计适合其主板(要求)的形状是很重要的。笔记本电脑的机箱通常是它的翻盖外壳，但是最近几年，它的外形已有很大差别，例如为适合自身(需要)变为平板形的可拆屏笔记本电脑已经出现。

电源

电源将交流电转变为低压直流电，为计算机内部部件提供电能。笔记本电脑通过内置蓄电池提供能量，通常可以工作数小时。

主板

主板是一个综合电路系统，它连接着电脑其它部件，包括中央处理器、随机存储器、磁盘驱动器(光盘、数字化视频光盘、硬盘或其它)，以及通过接口或膨胀槽连接的外围设备。

Passage B 计算机硬件(Ⅱ)

扩充插件板

在计算机信息处理技术中，扩充插件板是一个能被插进计算机主板或底板扩展槽的印刷电路板，并通过扩展总线，增加计算机系统的功能。扩充插件板可以用于获得或扩充主板不能提供的功能。

计算机数据存储器

储存设备是为储存端口和提取数字文件与目标(信息)而设置的计算机工作硬件和数字媒介。它既能临时又能永久地控制和存储信息，并能从内部或外部提供给计算机、服务器或任何类似电脑装置。数据存储是计算机的一个核心功能和基本组成。

虚拟硬盘

计算机使用各种媒介对数据加以储存。硬盘由于其高性能、低成本几乎使用于所有旧式计算机中。但相比之下固态硬盘更快更高效，尽管当前就美元而言每十亿字节比硬盘相对昂贵而因此出现在了2007年后组建的个人电脑中。一些系统可以使用磁盘阵列控制器从而实现其更高的性能和可靠性。

可移动媒介

为了实现计算机之间的数据转移，可以使用闪存盘和光盘。它们的有效性依赖于系统的可读性。大多数计算机有光盘驱动器，事实上，所有的计算机至少有一个U盘插孔。

输入和输出的外围设备

外围设备　输入和输出设备与计算机主机箱一样以外部封装为特点。下列(对此的描述)对许多计算机来说也许标准，也许很普通。

输入设备　输入设备允许用户将信息输入系统或控制它的运行。多数个人计算机有鼠标和键盘，但笔记本电脑其特色是以触屏代替鼠标。其它输入设备包括网络摄影机、麦克风、(计算机游戏的)操纵杆以及扫描仪等。

输出设备　输出设备以人们可阅览的形式提供信息。这些设备包括打印机、扬声器、显示器或盲人点字浮雕器等。

Unit Two　Computer Software

Passage A　计算机软件

计算机软件或简称软件，由数据和计算机指令组成，是计算机系统的一部分，其与系统的物理硬件形成对照。在计算机学和软件工程中，计算机软件由计算机处理的所有信息以及程序和数据组成。计算机软件包括计算机程序、文库以及相关的诸如在线文件或数字媒介等非执行数据。计算机硬件和软件互为需要，任何一方在实际操作中都不能独立应用。

在最低级状态中，执行代码包括特定机器语言指令的个体处理程序——代表性的是中央处理单元(CPU)。机器语言包括数个改变计算机之前状态的代表处理器程序指令的二进制数值组。例如，某一指令可以改变存储在计算机特定位置的数值——这个结果是使用者不能直接观察到的。一个指令也可以(非直接地)引起一些变化并使其出现在计算机显示屏上——对使用者来说这应该是可见的状态变化。处理器按照命令提供的先后顺序执行指令，除非它被指示“跳”到另一个不同的指令，或者被中断指令(到目前为止，多核心处理器具有统治地位，其中每一个“核心”按顺序运行指令，但不管怎样，每个应用软件根据系统默认值仅仅在一个“核心”运行，但是有些软件被设定成多核心运行)。

大多数的软件由高级程序语言编写而成，对于程序员来说，因为它们比起机器语言更接近自然语言而更容易、更有效。高级程序语言通过编译器、或者注释器、或者二者的联合被转换成机器语言。软件也可以由低级的汇编语言写成，它与计算机的机器语言指令非常一致，并使用汇编程序转换成机器语言。

Passage B　应用软件与系统软件

在所有实际计算机平台中，软件可以被分为几大类。就使用目的来说，计算机软件可以分为应用软件和系统软件。

应用软件

应用软件是利用计算机系统超越计算机本身的基础操作完成一些特殊功能或者提供娱乐功能的软件。由于现代计算机要“执行”的任务范围庞大，所以计算机应用软件种类众多。

系统软件

系统软件是直接运行计算机硬件从而提供给用户和其它软件需要的基础功能、并同时为运行应用软件提供平台的软件。系统软件包括要素如下：

操作系统

操作系统是软件的基础组合，它管理资源并为其它优先运行的软件提供常规服务。监督程序、引导装载程序、框架系统和窗口系统是计算机操作系统的核心。实际上，为便于用户使用仅有一个操作系统的计算机来做更多可能性的工作，操作系统可自带附加软件(包括应用软件)。

设备驱动程序

设备驱动程序操作着或控制着附加在计算机上的一种特别设备。每个设备至少需要一个相应的设备驱动程序。因为典型的计算机最低限度至少要有一个输入设备和一个输出设备，所以一个典型的计算机需要一个以上的设备驱动程序。

实用程序

实用程序是设计出来用于用户维护和保护他们的计算机的程序。

恶意软件

恶意软件是用于破坏和干扰计算机工作的软件。就其本身而言，恶意软件是不受欢迎的。恶意软件与计算机犯罪紧密相关，尽管有些恶意软件可能是作为无伤大雅的笑话而被设计出来的。

Unit Three Computer Network

Passage A 计算机网络

计算机网络或数据网络是一个允许节点共享资源的数字通信网络。在计算机网络中，联网的计算设备利用数据传输器互相交换数据。节点之间可以通过电缆媒体或无线媒体建立联系。

网络节点

网络节点指与网络连接的电脑或其它设备，它们拥有独立地址，并能够发送和接收数据。节点可以包括主机，例如个人电脑，电话，服务器和网络硬件。不管是否是直接连接，当其中一个设备能够与其它设备交换信息时，这两种设备就可看做联网了。

网络协议

为了使数据传输运行起来，你需要用协议来管理数据如何被网络设备发送、接收和翻译。

协议在几个层面堆积起来，形成互联网协议组。最底层(连接层)是数字电子技术处理协议。在局域网，这个层面被称作以太网协议。上面是网络层，在该层中有互联网协议(IP)。从这儿开始，一个被称为IP地址的概念就出现了。接下来，在传输层，我们发现了传输控制协议，简称TCP，和用户数据报协议(UDP)。而且，还可能会有其它协议。再往上一层是应用层，这里有我们从日常使用中了解的协议，如超文本传输协议、简单邮件传输协议，以及点对点语音通信协议。

计算机网络支持大量的应用和服务，如万维网，数字视频，数字音频的访问，应用程序和存储服务器、打印机、传真机的共享使用，电子邮件和即时通信应用软件的使用等等。计算机网络在传输信号的媒介、管理网络流量的通信协议、网络规模、拓扑和组织目的等方面会有所不同。最著名的计算机网络是因特网。

Passage B 网络拓扑

网络拓扑是指通讯网络中各种设备的布局，包括它的节点和连接线。有两种方式定义网络几何结构:物理拓扑和逻辑(信号)拓扑。

网络的物理拓扑是工作站的实际几何布局。下图中展示了几种常见的物理拓扑类型。

在总线型网络拓扑结构中，每一个工作站都与总线连接。所以，实际上每一个工作站都被直接连接到网络中的其它工作站上。

在星形网络拓扑结构中，其它入网机器直接连接到一台中心处理机或服务器上，每一台工作站都要通过中心处理机或服务器才能与其它工作站间接连通。

在环形网络拓扑结构中，所有工作站都被连接成一个封闭的环形结构。相邻的两个工作站直接连接在一起。其它工作站间接相连，数据通过一个或更多的中间节点进行传输。

如果在星形网络拓扑和环形网络拓扑使用令牌环网的话，信号只能是单向传输，通过所谓的令牌从一个节点传输到另一个节点。

网状拓扑结构使用全网状拓扑和部分网状拓扑两种方案。在全网状拓扑中，每一个工作站都直接与其它工作站相连。在部分网状拓扑中，有些工作站与所有其它工作站相连，有些只是与那些数据交换最频繁的节点相连。

树形拓扑结构会使用相互连接的两个或更多的星型网络。星型网络的主机被连接到总线。因此，树形网络是星形网络的总线型网络。

逻辑(或信号)拓扑指的是信号从一个节点到另一个节点的传输路径的特性。在许多情况下，逻辑拓扑与物理拓扑无异。但是情况并非一直如此。例如，一些网络以星型结构的物理拓扑布局，却以总线或环状网络的逻辑拓扑进行操作。

Unit Four　Multimedia

Passage A　多媒体

什么是多媒体

多媒体是不同内容形式的综合体，包括文本、音频、图像、动画、视频和互动性内容等多种媒体形式。英文学习网站就是一个完美的多媒体示例：网站上有关于知识点的文本，有关于文章的音频材料，甚至还有在真实情境下演绎学习内容的视频作品。

多媒体应用

多媒体应用于各个领域，包括但不限于广告、艺术、教育、娱乐、工程、医学、商业和科学研究。几个例子如下：

商业用途

商业广告艺术家和平面设计师所采用的许多新旧电子媒体都是多媒体。在广告中，激动人心的展现形式被用来抓住并保持观众的注意力。企业之间以及办公室之间的交流通常也会通过创意服务公司来实现，以此来寻求优于简单幻灯片播放的高级多媒体演示，用于销售理念和活跃培训。商业多媒体开发者还可以为政府服务或进行公益服务应用设计。

娱乐

多媒体在娱乐产业应用广泛，尤其是在电影和动画的特技效果开发上。多媒体游戏也是非常流行的消遣方式，它们是既可使用光盘又可在线直接运行的软件程序。一些电子游戏也具有多媒体特征，用户从坐在一边的被动信息接受者转变为活跃的参与者的多媒体应用称为交互式多媒体。在艺术领域，有多媒体艺术家，他们能够使用不同媒体把想法与技术相融合，在某种程度上，实现了与观众的互动。相关性最高的当属把电影与歌剧和各种数字媒体融合到一起的彼得·格林纳威了。

教育

在教育行业，多媒体被用来制作计算机辅助培训课程(通常被称 CBTs)和撰写类似百科全书和年历这样的参考书。用户通过计算机辅助培训完成关于特定主题的一系列的报告、文本和各种信息格式的相关图表。寓教于乐的产品就是教育和娱乐的结合，尤其是多媒体娱乐的结合体。

多媒体的未来

在科幻小说《神经漫游者》中，威廉·吉布森描述了一个空间，由计算机控制的网络空间。一旦大脑与电脑相连，人就会感受这个空间的所有经历。他在现实世界的各种感受也会被一系列新的电子刺激所代替。这个网络空间就是将来虚拟现实技术的目标。虚拟现实将触感同视频和音频媒体结合起来，使玩家置身于虚拟世界。例如，在虚拟空间，学生可以解剖人的身体，游览古战场，甚至和莎士比亚交谈。

Passage B 如何创建多媒体演示文稿

多媒体演示文稿与普通演示文稿的不同之处在于它包含一些媒体或动画的格式。典型的多媒体演示文稿至少要包含以下的一种元素：

视频或影片剪辑

动画

声音(可能是画外音、背景音乐或声音片段)

导航结构

多媒体演示文稿技术的选择

首先要做的也是最难的是选择制作演示文稿的技术。归根结底是在两个主要的竞争者之间进行选择：Adobe Flash 或者 Microsoft PowerPoint.

Adobe Flash

Flash 使你能够在演示文稿中插入令人震撼的动画。同时，它还具有非常过硬的视频压缩技术。

可能 Flash 最大的优势在于它能够把演示文稿直接上传到你的网站。

但是 Flash 最大的问题是它是一个非常难掌控的系统。我曾经参加过一个培训课程，有幸接触到几个平面设计师可以寻求帮助，但仍然感觉用 Flash 完成演示文稿是件很难的事情。

最新版本(Flash 8 and Flash CS3)的操作简单了许多。新版本有一种新功能叫做 Flash 幻灯片演示文稿。与传统的时间轴不同，它允许用户创建或者是插入幻灯片，有点像 PowerPoint 的幻灯片排序。

新版本仍然很贵。亚马逊上 Flash (CS3)的最新版本会让你破费 515 英镑(626.99 美元)。你很可能还要把上一个像样的 Flash 培训班所需的约 400 英镑考虑进来。

有一种 Flash 的低成本替代工具叫做 Swish. 它们不需要详细的编程知识，使创建 Flash 演示文稿变得简单起来。你可以登录 www.swishzone.com 网站获取免费试用软件。

Microsoft PowerPoint

最简单的制作多媒体演示文稿的方法是使用 Microsoft PowerPoint。你可以插入视频、配乐甚至是动画。

到目前为止，使用 PowerPoint 制作多媒体演示文稿的最大好处就是任何人都能轻松地编辑演示内容。

介绍演示文稿中多媒体效果的最好例子就是右图的出租车序列。屏幕上最初显示一幅线条画，紧接着一辆出租车逐渐显现出来。这个序列只是 PowerPoint 里一个模板的其中一部分。你可以通过在下载页下载出租车动画模板看到运行效果。确保你要在 Power Point 放映模式下观看这个序列。

在 Presentation Helper 网站上，有很多教程可以帮助你更轻松地组合多媒体演示文稿，如：

怎样在多媒体演示文稿中插入视频剪辑

怎样在 PowerPoint 中插入音乐

PowerPoint 简单菜单

怎样让你的幻灯片动起来

把这些元素结合起来，你就会找到在 PowerPoint 中做多媒体展示文稿的方法。

Unit Five　E-commerce

Passage A　电子商务

电子商务是指在互联网上进行的商品和服务的买卖，交易双方无需见面。

众所周知，消费者、企业和政府是电子交易和业务流程的主要参与者。据此，电子商务一般被划分为以下几种类型：企业间电子商务、企业对消费者电子商务、消费者间电子商务、企业与政府间电子商务以及政府间电子商务。其中，企业间电子商务、企业对消费者电子商务、消费者间电子商务是最常见的三种商务模式。

企业间电子商务

企业间电子商务模式是指产品或服务在企业之间而非企业与最终用户之间的交易。这类交易可以发生在企业与供应链成员之间或任何其它企业之间。例如，家具制造商需要诸如木材、油漆、清漆等原材料。在 B2B 商务模式中，它就可以向供应商电子下单并多次电子支付。

企业对消费者电子商务

企业对消费者电子商务模式指企业通过网络直接将商品销售给客户的过程。在这种模式中，企业和个人客户是主要参与者。客户浏览你方网站上的产品信息页，选择产品，并且交货前在结账区选择信用卡、借记卡或其它电子支付方式进行支付。当然，客户需要输入详细地址，选择你提供的一种运输方式。基本的企业与客户间电子商务体系是相对简单的。你需要在网站上展示产品和价格，记录客户的详细信息并设立专门接受付款的结账区。

消费者间电子商务

消费者间电子商务指消费者之间进行的交易。自从社交网络出现后，C2C 成为电子商务中增长最快的一种模式。有很多网站免费提供分类广告、拍卖和论坛服务，供个人用户进行商品交易。1995年成立的易趣购物平台的拍卖服务就是该种电子商务最好的例子。同时，多亏了网上支付系统如贝宝的出现，人们可以毫不费力地网上支付和收款。

企业与政府间电子商务

在这种模式中，政府扮演的是电子商务用户的角色，他们向企业购货，同时执行宏观管理职能来支持和引导电子商务。

政府间电子商务

政府间电子商务指的是政府各部门之间的电子商务活动。在这个模式中，大部分的政府活动可以通过高速、高效率和低成本的网络技术来展开。

Passage B 怎样优化电子商务网站?

电子商务世界变幻莫测，吸引顾客变得越来越难。大量的在线服务供应商为吸引顾客的注意而绞尽脑汁。问题是怎样才能使网站在网上可见并出现在潜在客户面前呢?

在优化电子商务网站的过程中，世界各地的搜索引擎优化专业人员和咨询公司会专注于以下四个关键领域来提高销售业绩。

产品说明

产品说明不仅要有相关性，还要详尽细致。潜在客户会首先阅读产品说明。在了解产品或服务之后，如果他们发现产品是有用且有趣的，才会决定购买。

产品说明还应该是独特的。如果你使用其它网站的任何内容，会被谷歌认定为复制内容，你的网站也会因此而受到处罚。

长尾关键词

长尾关键词是指比一般搜索的关键词更详细、通常也更长的关键词或短语。因为更加具体，长尾关键词虽然只占用较少的搜索流量，却拥有更大的转化价值。

举个例子来说，如果你是一家销售古典家具的公司，只搜索“家具”的话，你的网页永远都不可能出现在自动搜索的前列，因为竞争对手实在是太多了。但是如果你专营，例如，当代装饰艺术家具，那么类似“当代装饰艺术风半圆躺椅”的关键词会更靠谱地寻找到那些正好搜索这类产品的客户。

产品评论

大约百分之七十的购买者在购买商品之前都会搜索产品信息。顾客相信其他人的观点和反馈。确保给顾客提供写评论的可能性。这样的话，潜在客户不用离开你的页面就会得到关于产品质量的客观反馈。产品评论的另外一个好处是它是通过新颖、独特的内容大大提高搜索引擎优化价值的免费内容。

结构化数据和富摘要

结构化数据是指搜索结果中被编码在统一资源定位器地址的几行搜索结果。不同的结构会使搜索结果更加吸引人，并且与竞争者区分开来。

电子商务富摘要包括:星、等级、评论数量、图像和价格区间。富摘要旨在为链接内容提供更好的回顾。它能够使用户在进入网页之前作出决定——搜索结果是否满足他们的疑问。

电子商务跟踪

不断增加的访问者数量并不意味着你可以将“武器”束之高阁，吃老本。想了解关键词带来的真正收益和产品销售收入，你需要跟踪客户的销售路径。能够监视网上商城的谷歌分析是个不错的选择。

Unit Six　Computer Security

Passage A　计算机安全

生活充满了权衡，计算机安全也不例外。绝对安全意味着形同虚设，至少是孤立的。没有人接近它，也就没有人能够破坏它。唯一的问题是，在这种状态下，它亦是无用的计算机。所以，计算机安全的程度总是取决于在使用它和限制其误用之间的权衡。你需要对确保计算机安全所花费的时间、金钱与它一旦被侵入或破坏所造成的损失来进行权衡。换句话说，你是不会浪费锁头和钥匙来保证废物的安全的。

国际标准化委员会对计算机安全给出的定义是：为数据处理系统所建立和采取的技术以及管理的安全保护，保护计算机硬件、软件、数据不因偶然的或恶意的原因而遭到破坏、更改、泄露。计算机安全包括实体安全、软件安全、数据安全和操作安全等。

实体安全指计算机系统的设备和相关装置有序运行，包括主机、网络设备、通信线路、存储设备等。

软件安全指软件的完整性。换句话说，即操作系统软件、数据库管理软件、网络软件以及相关应用及材料的完整性，包括软件的开发、软件安全测试、软件修改及复制等。

数据安全指系统拥有或创建的数据或信息在不被毁坏和泄露的情况下被充分、有效、合法地使用，包括输入、输出、用户识别，访问控制，备份和修复等。

操作安全指合法使用系统资源，包括电源管理、环境(包含空调)、人员、机房出入、数据和媒介管理、操作管理等。

当你设计和修改计算机和网络安全的时候，要考虑清楚：你想怎样使用系统？如果安全不复存在，你将失去什么？它会帮助你选择解决方案，了解相关的复杂性及其成本。

Passage B　一些计算机安全技术

计算机安全与信息、硬件和软件的保护有关。安全措施主要包括加密、限制访问、预防灾难和制作备份。

加密

当信息通过网络发送时，总是存在越权访问的可能性。信息传输的距离越远，安全风险就越大。例如，局域网上的电子邮件只会遭遇在办公室等受控环境中进行操作的有限数量的用户。而在国家信息高速公路上穿梭在全国的电子邮件，则提供了更多的被截取的机会。政府鼓励使用国家信息高速公路的工商企业采用一种特殊的加密程序。这种程序被存储在一种处理器芯片上，该芯片被称为加密芯片，通常也叫密钥托管芯片。个人也可使用加密程序来保护其私人通信。使用最广泛的加密程序之一是 PGP 加密软件。

限制访问

安全专家不断设计新方法，保护计算机系统免受未经授权之人的访问。有时，安全措施就是派警卫守卫公司计算机室，检查每个进入者的身份证明。更多时候，它是将密码谨慎地分配给员工，并且在他们离职时加以更换。

今天，大多数大公司都使用被称为防火墙的特殊硬件和软件，来控制对其内部计算机网络的访问。这些防火墙，在公司专用网络与包括因特网在内的所有外部网络之间，起到安全缓冲区的作用。所有进出公司的电子通信都必须经过防火墙的评估。通过拒绝未经授权通信的出入，来维护安全。

预防灾难

不做灾难预防的公司(甚至个人)是不明智的。大多数大机构都有灾难恢复计划。硬件可以锁起来，但雇员常常觉得这种约束很麻烦，因此安全措施也就松弛了。火与水可能对设备造成极大的损害。因此，许多公司会与其他公司达成合作协议，在出现灾难的情况下分享设备。

备份数据

设备可以随时更换，公司的数据却可能无可替代。大多数公司首先会采取一些方法来防止软件和数据被篡改。这些方法包括仔细审查求职者，严守密码，以及时常检查数据和程序。然而，最保险的办法还是经常制作数据备份，并将其存放在远端地点。

Unit Seven Big Data

Passage A 大数据(Ⅰ)

大数据是为描述数据集合而生成的术语，数据集数量巨大复杂，传统数据处理应用软件不能处理它们。大数据挑战包括捕捉数据、数据存储、数据分析、调查、分享、转换、可视化、查询、更新和信息隐私。

最近，大数据更倾向于指预测分析和用户行为分析的应用，或者指从数据抽取价值以及从几乎不能达到特别规模的数据集抽取价值的某一高阶数据分析方法。“毫无疑问，当前可获得的数据的量是相当巨大的，但它不是这个新数据系统最具关联的特征。”数据集的分析能够找到“发现市场趋势、预防疾病、打击犯罪等等”的新的相关性。科学工作者、业务主管、医学从业者、广告业、政府等经常遇到大数据集所涉及的包括因特网搜索、金融科技、城市信息学和商业信息学等领域带来的麻烦。科学工作者在电子科研领域受到限制，其中包括气象学、基因组学、关系学、复合物理仿真、生物学和环境研究。

数据集的迅速发展某种程度上是由于它们越来越多地被价格低廉的以及数量众多的信

息传感物联网设备所收集，诸如移动设备、遥感、软件登录、摄像机、麦克风、无线射频识别器和无线传感网络等。世界上人均科技信息储量自20世纪80年代以来每40个月增长一倍。到2012年，每天产生2.5艾字节(2.5×10^{18})的数据。对于大企业来说决定谁应当在影响整个组织的大数据上拥有主动权是一个大问题。

相关的数据库管理系统和桌面统计以及可视化打包在处理大数据时经常遇到困难。此项工作可能需要大量的平行软件运行于数十、数百、甚至数千的服务器上。大数据的价值是:大数据改变了用户对权限和他们的工具的依赖，并扩充权限使大数据成为一个变动指标。“对于一些组织来说，面对数百千兆字节的数据，第一时间可能激起一种‘重新考虑数据管理选择’的需求。(而)对于其它组织来说，在数据规模变成重大关注点之前，可以提取其数十、数百兆兆字节。”

Passage B　大数据(Ⅱ)

大数据，这个术语自20世纪90年代已开始使用。大数据通常涉及数据集，其规模超越平常使用的软件工具捕获、注意、管理以及在一个容许错过的时间处理数据的能力。大数据哲学包括无组织的、半组织化的和有组织的数据，但主要焦点在无组织的数据上。大数据的“尺寸”是一个不断变动的指标，到2012年其数据规模已从几十兆兆字节达到数千兆字节。大数据需要一套技巧和工艺，这套技巧和工艺以整合的新形式显示来自于数据集多样的、复杂的、超大规模的洞察力。

2001年，在研究报告和相关讲座中META集团(现Gartner公司)以增加量(数据量)、速度(数据输入输出的速度)和多样性(数据类型和资源的范围)三个维度来定义数据增长的挑战和机遇。Gartner公司，以及大部分的行业现在仍继续使用这个“3V”模式来描述大数据。2012年，Gartner公司更新了对它的定义如下:“大数据是大容量、高速度和(或)超多样性的信息资产，它需要划算的、新颖的信息处理形式来加强洞察力、决策和过程自动化。”Gartner的“3V”定义仍被广泛应用着，与其相一致的定义这样叙述:“大数据代表了对于它的价值转换需要特别的技术和分析方法的大容量、高速度和(或)超多样性为特征的信息资产。”另外，一个新V即“精确性”被一些组织加进其定义中，被一些行业权威列为修正性挑战。“3V”现在已经扩充了大数据其它的补充特性:

- 数据量:大数据没有样本；它仅仅观察和追踪所发生的事物；
- 速度:大数据经常实时提供；
- 多样性:大数据来自于文本、图像、声音和视频，另外它通过数据融合修复失踪数据；
- 机器学习能力:大数据经常不问为什么而是简单地检测模式；
- 数字足迹:大数据经常是数字化交互的一个免费副产品；

(这个概念日益成熟并更鲜明地描述了大数据和商业智能之间的不同。)

- 商业智能对资料使用描述性统计，以数据和高信息密度测量事物、发现趋势等；
- 大数据使用来自于非线性系统识别的归纳性统计和概念，以推出法则(回归、非线性关系和因果效应)，而这些法则来自于低信息密度、揭露关联和相关性或者执行成果和行为预报的巨量数据集。

Unit Eight Cloud Computing

Passage A 云计算

云计算是一种因特网计算的形式，它给计算机及其它需要的设备提供可共享的计算机处理的资源和数据。它是一种模式:可以无处不在，按需进入可配置的计算资源分享池(例如计算机网络、服务器、存储、应用和服务)，只要付出较少的管理工作就可以准备和发出所需的资源。从根本上说，为了使数据获取机制更容易更可靠，云计算允许拥有多方面能力的用户和企业以私人拥有的云、或者以第三方服务器处理和加工他们的数据。数据中心可以设置在距离使用者很远的地方，可以跨越城市，甚至跨越地球。云计算依靠资源共享实现一致性和规模经济，这一点与通用电网的效能相似。

云计算的拥护者声称云计算可以使公司避免先期基础设施投资(例如购买服务器)。同样地，它能使组织机构把精力集中在他们的核心业务上，而不是花费时间和金钱用于计算机基础设施建设。支持者还声称，通过改善管控性和减少维修以及使信息技术团队更快速地调整资源以满足波动的、不可预知的商业需求，云计算可以使企业更快地启动和运行其应用程序。云供应商通常采用“账单到期即付”模式。如果管理员不适应云定价模式，将产生不可预期的高额费用。

2009 年，大容量网络、低成本计算机和存储设备的使用，以及硬件虚拟化、服务导向构架，自主实用的计算的广泛采用大大促进了云计算的发展。当云计算需要增加，则公司则按比例增加，反之则减少。据报道，在 2013 年，由于云计算的高计算能力、低成本服务、高效率性、可扩展性、可达性以及可用性，使它成为了一个被高度需求的服务与应用。一些云供应商每年的业务增长率达到 50%。当然云计算还处在初期阶段，它还有许多需要解决的缺陷，才能使运算服务更可靠，让用户更满意。

Passage B 云计算的主要特征

- 因为云计算可能会增强用户在重新配置、添加和扩大技术基础设施资源时的灵活性，所以组织的灵活性会得到提高。
- 云供应商声称成本减少。公共云交付模型把资本支出(如购买服务器)转换成业务支出。据称这一环节降低了入口阻碍，因为基础设施通常由第三方提供，不必为了一次性的或并不频繁出现的计算任务而买入。基于应用计算的定价非常细化，提供根据使用来选择的计费方式。
- 设备与位置独立性使用户不管身在何处，使用何种设备(例如个人计算机、移动电话)都能够使用网络浏览器访问系统。因为基础设施在异地(通常由第三方提供)并经由因特网访问，所以用户可以从任何地方与它连接。
- 云计算应用软件的维修更容易，因为它们不需要安装在每个用户的计算机中并且可以从不同的地点进入(例如不同的工作地点，外出旅游地等等)。

• 多租户技术使多名用户共享资源，并共同分担成本。

• 效能性由服务供应商的IT专家监控。用互联网服务作为系统界面来构架具有一致性和松散耦合性的构造。

• 当多个用户可以同一时间使用同一数据，而不需要等待这一数据被保存或通过电邮被发送时，生产率提高了。

• 众多过剩网站的启用提升了可靠性，也使精心设计的云计算适合交易的连续性和灾难的恢复。

• 可扩展性和灵活性。通过动态的(按需)资源供给，以精细的、自助的方式为基础，用接近实时的原则(工作)提供服务，而不需要用户处理高峰时段下载。这就使其拥有了“当需求增加就按比例增加、如果资源不被采用就下降”的能力。

• 由于数据集中管理同时增加“资源安全集中化”，(云计算)在安全上能够得到改善。云计算的安全性与其它传统系统一样好，甚至比其它传统系统更好，在某种程度上因为服务供应商能够致力于研究解决许多客户无力处理的或者缺乏解决的技术的安全问题。然而，当数据分布在一个广大的区域或大量的设备上，以及在多租户系统中由无关的用户共享时，安全的复杂性也在迅速增加。另外，用户想要进入安全日志可能是很困难的或者是不可能的。私有云的安装从某种程度上来说是由于用户“保留控制基础设施、避免失去对信息安全控制”的(私人)愿望。

Appendix C Keys to the Tasks

Unit One Computer Hardware

Section One Speaking

Open.

Section Two Listening

Task Ⅱ

1. Listen carefully and choose the words you hear.

(1) C (2) B (3) B (4) D (5) D (6) A (7) D (8) C (9) A (10) B

2. Listen to the short passage and choose the proper words to fill in the blanks.

(1) British (2) greeting (3) know (4) Pleased (5) day

3. Listen to the dialogues and choose the best responses to what you hear.

(1) D (2) D (3) B (4) C (5) A (6) C (7) A (8) B

4. Listen to the dialogues and choose the best answer to each question you hear.

(1) C (2) D (3) A (4) B (5) B

5. Listen to the dialogue and answer the following questions by filling in the blanks.

(1) at the airport (2) a manager (3) bad (4) a long way

(5) nine

6. Listen to the passage and fill in the blanks with words you hear. Some new words given below will be of some help to you.

(1) interest (2) friends (3) classes (4) English (5) beach

(6) picnic (7) parties (8) friendship (9) fresh (10) meet

Section Three Reading

Passage A

Task Ⅲ - 1

1. Fill in the blanks without referring to the passage.

(1) collection (2) system (3) components (4) keyboard

(5) software (6) stored

2. Answer the following questions according to the passage.

Open.

3. Complete each of the following statements according to the passage.

(1) the software (2) a combination

(3) the form factor (4) tablet computers

(5) a built-in battery (6) the other parts

4. Translate the following sentences into English.

(1) I put some of my things in storage.

(2) When you look at their new system, ours seems very old-fashioned by contrast.

(3) Peter decided to contact the manager directly.

(4) I like drinks such as tea and soda.

(5) The company deals in both hardware and software.

(6) The rocks were formed more than 4,000 million years ago.

(7) Indeed the personal computer helped drive a new era in both science and business.

(8) You can either have tea or coffee.

5. Translate the following sentences into Chinese.

Omit.

Passage B

Task Ⅲ - 2

1. Match the following English phrases in Column A with their Chinese equivalents in Column B.

(1) d (2) a (3) g (4) f (5) i (6) h (7) e (8) j (9) b (10) c

2. Decide whether the following statements are T(true) or F (false) according to the passage.

(1) F (2) F (3) T (4) T (5) F (6) T (7) T (8) F (9) T (10) T

3. Translate the following passage into Chinese.

Omit.

Extra Reading

Read the following passage and choose the best answer to each question.

1. B 2. D 3. D 4. C 5. A

Section Four Writing

Task Ⅳ

1. You are required to write a Telephone Message according to the following instructions given in Chinese.

[参考范文]

To: Mr. William

Time: 10 a. m., June 25th, 2016

Message: Mr. Black of Honeywell Co. is going to Beijing on business tomorrow. His appointment with you on the morning of the 27th has to be cancelled. Mr. Black will contact you for another appointment when he is back.

From: Jane(secretary)

2. You are required to make a Name Card according to the following instructions given in Chinese

［参考范文］

Foreign Affairs Department
Shenyang International School

Wang Jiawei
Director/Professor

Address: 666 Chongshan Road, Huanggu, Shenyang
Tel & Fax: 024-22235555E-mail: gaop702@163. com
Post code: 110102

Unit Two Computer Software

Section One Speaking

Open.

Section Two Listening

Task Ⅱ

1. Listen carefully and choose the words you hear.

(1) C (2) D (3) B (4) D (5) A (6) C (7) C (8) A (9) B (10) C

2. Listen to the short passage and choose the proper words to fill in the blanks.

(1) pretty (2) always (3) school (4) English (5) music

3. Listen to the dialogues and choose the best responses to what you hear.

(1) A (2) B (3) D (4) C (5) C

4. Listen to the dialogues and choose the best answer to each question you hear.

(1) A (2) B (3) C (4) C (5) B

5. Listen to the dialogue and answer the following questions by filling in the blanks.

(1) Li Hua　(2) a lot　(3) long hair　(4) are still shortsighted
(5) Jerry

6. Listen to the passage and fill in the blanks with words you hear. Some new words given below will be of some help to you.

(1) winter　(2) took　(3) put　(4) around　(5) eat
(6) drink　(7) buy　(8) But　(9) lady　(10) share

Section Three Reading

Passage A

Task Ⅲ - 1

1. Fill in the blanks without referring to the passage.

(1) computer instructions　(2) information　(3) data
(4) libraries　(5) digital media　(6) each other

2. Answer the following questions according to the passage.

Open.

3. Complete each of the following statements according to the passage.

(1) data
(2) be realistically used
(3) the value
(4) default
(5) natural languages
(6) a compiler

4. Translate the following sentences into English.

(1) They are familiar with the opinion that all matter consists of atoms.
(2) Ican help you with the new computer program.
(3) He looks healthier in contrast to his former self.
(4) There is no executable code at this location in the source code.
(5) We need to carry out our plans.
(6) Application software performs specific tasks for the user.
(7) We would all be programming in assembly language or even machine language.
(8) It seems to serve no visible use.

5. Translate the following sentences into Chinese.

Omit.

Passage B

Task Ⅲ - 2

1. Match the following English phrases in Column A with their Chinese equivalents in Column B.

(1) d　(2) c　(3) b　(4) a　(5) g　(6) e (7) j　(8) h　(9) f　(10) i

2. Decide whether the following statements are T(true) or F (false) according to the passage.

(1) F　(2) F　(3) T　(4) F　(5) T　(6) F　(7) T　(8) T　(9) F　(10) T

3. Translate the following passage into Chinese.

Omit.

Extra Reading

Read the following passage and choose the best answer to each question.

1. C　2. D　3. A　4. C　5. A

Section Four　Writing

Task Ⅳ

1. You are required to complete the resume according to the information given.

［参考范文］

Test Preparation Course Application Form

Family Name: Zhang　　First Names: Jinqiu

Date of Birth: Oct. 16, 1982　　Nationality: Chinese

Sex: Female　　Telephone Number: 000-67289879

Address: 496 Hongqiao Road, Yanjing City

I am a sophomore in Liaoning University. I am majoring in international trade. Though I have little difficulty in reading English materials, I am totally at a loss when I attend my American teacher's lecture. I'm eager to improve my listening ability in your seminar. I look forward to hearing from you soon.

2. You are required to complete the resume according to the information given.

［参考范文］

Resume

Name: Li Aihua

Address: 61 Fixing Road, Beijing

Date Of birth: May 16, 980

Sex: Male

Marital Status: single

Job objective: Seeking a job as a computer programmer

Education: I graduated from the Computer Department of Huaguang Technical College in 2001 with excellent scores.

Foreign languages: I studied English during the 3 years at college. I'm excellent at reading. I can communicate in English fluently.

Hobbies: I like swimming and surfing the Internet.

Unit Three Computer Network

Section One Speaking

Open.

Section Two Listening

Task Ⅱ

1. Listen carefully and choose the words you hear.

(1) A (2) B (3) D (4) C (5) B (6) D (7) B (8) C (9) C (10) A

2. Listen to the short passage and choose the proper words to fill in the blanks.

(1) standard (2) 80% (3) billion (4) digital (5) Digital

3. Listen to the dialogues and choose the best responses to what you hear.

(1) A (2) B (3) D (4) D (5) B (6) A (7) C (8) D

4. Listen to the dialogues and choose the best answer to each question you hear.

(1) B (2) C (3) A (4) B (5) C

5. Listen to the dialogue and answer the following questions by filling in the blanks.

(1) swimming player (2) 20 (3) at the age of 15 (4) she is tired of it

(5) at 6 a.m.

6. Listen to the passage and fill in the blanks with words you hear. Some new words given below will be of some help to you.

(1) ability (2) communicate (3) wrong (4) recognize (5) sound

(6) sights (7) sounds (8) upset (9) immediately

(10) overloaded

Section Three Reading

Passage A

Task Ⅲ - 1

1. Fill in the blanks without referring to the passage.

(1) nodes (2) resources (3) data (4) data link

(5) cable media (6) wireless media

2. Answer the following questions according to the passage.

Open.

3. Complete each of the following statements according to the passage.

(1) nodes

(2) data link

(3) network nodes

(4) protocols

(5) digital electronics

(6) the transmission medium, communication protocols, the network sign, topology & organizational internet.

4. Translate the following sentences into English.

(1) The Internet is a worldwide computer network.

(2) We think that she is absolutely capable of taking care of herself.

(3) Money is a medium for buying and selling.

(4) He plays the guitar as well as you.

(5) Women officers make up 13 percent of the police force.

(6) The only thing is no wireless devices.

(7) The only access to the town is across the bridge.

(8) People differ in their attitudes towards failure.

5. Translate the following sentences into Chinese.

Omit.

Passage B

Task Ⅲ - 2

1. Match the following English phrases in Column A with their Chinese equivalents in Column B.

(1) g　(2) j　(3) e　(4) c　(5) i　(6) f　(7) d　(8) a　(9) b　(10) h

2. Decide whether the following statements are T(true) or F (false) according to the passage.

(1) F　(2) T　(3) F　(4) T　(5) F　(6) T　(7) T　(8) T　(9) F　(10) F

3. Translate the following passage into Chinese.

Omit.

Extra Reading

Read the following passage and choose the best answer to each question.

(1) B　(2) D　(3) A　(4) A　(5) C

Section Four　Writing

Task Ⅳ

1. You are required to write a Telephone Message according to the following instructions given in Chinese.

[参考范文]

CONTRACT

Date: July 25, 2015　　　Contract No.: VH756778

The Buyers: DL Company

The Sellers: CBD Company

This contract is made by and between the Buyers and the Sellers; whereby the Buyers agree to buy and the Sellers agree to sell the under-mentioned goods subject to the terms and conditions as stipulated hereinafter:

(1) Name of Commodity: corn

(2) Quantity: 20000 tons

(3) Unit price: $4/kilo

(4) Total Value: $80, 000, 000

(5) Packing: the packing should be preventive from dampness and shock, and shall be suitable for ocean transportation.

(6) Country of Origin: U. S. A.

(7) Terms of Payment: L/C

(8) Insurance: insurance shall be covered by the seller for 110% of the invoice value against All Risks.

(9) Time of Shipment: by April, 10

(10) Port of Lading: New York, America

(11) Port of Destination: Dalian, China

(12) Claims:

Within 45 days after the arrival of the goods at the destination, should the quality, specifications or quantity be found not in conformity with the stipulations of the contract except those claims for which the insurance company or the owners of the vessel are liable, the Buyers shall have the right on the strength of the inspection certificate issued by the C. C. I. C and the relative documents to claim for compensation to the Sellers.

(13) Force Majeure

The sellers shall not be responsible for the delay in shipment or non-delivery of the goods due to Force Majeure, which might occur during the process of manufacturing or in the course of loading or transit. The sellers shall advise the Buyers immediately of the occurrence mentioned above the within fourteen days there after. The Sellers shall send by airmail to the Buyers for their acceptance a certificate of the accident. Under such circumstances the Sellers, however, are still under the obligation to take all necessary measures to hasten the delivery of the goods.

(14) Arbitration

All disputes in connection with the execution of this Contract shall be settled friendly through negotiation. In case no settlement can be reached, the case then may be submitted for arbitration to the Arbitration Commission of the China Council for the Promotion of International Trade in accordance with the Provisional Rules of Procedure promulgated by the said Arbitration Commission . The Arbitration committee shall be final and binding upon both parties, and the Arbitration fee shall be borne by the losing parties.

The Buyers:　　　　　　The Sellers:

Unit Four Multimedia

Section One Speaking

Open.

Section Two Listening

Task Ⅱ

1. **Listen carefully and choose the words you hear.**

   (1) B (2) D (3) D (4) C (5) B (6) B (7) D (8) D (9) D (10) B

2. **Listen to the short passage and choose the proper words to fill in the blanks.**

   (1) birthday (2) families and friends (3) sandwiches
   (4) Opening (5) wish

3. **Listen to the dialogues and choose the best responses to what you hear.**

   (1) B (2) C (3) D (4) A (5) C (6) B (7) A (8) C

4. **Listen to the dialogues and choose the best answer to each question you hear.**

   (1) D (2) B (3) B (4) A (5) B

5. **Listen to the dialogue and answer the following questions by filling in the blanks.**

   (1) restaurant (2) wine (3) waiter and guest (4) steak (5) salad

6. **Listen to the passage and fill in the blanks with words you hear. Some new words given below will be of some help to you.**

   (1) habits (2) freshly-cooked (3) vegetables (4) soup (5) salad
   (6) potato chips (7) knife (8) fork (9) chopsticks (10) rude

Section Three Reading

Passage A

Task Ⅲ - 1

1. **Fill in the blanks without referring to the passage.**

   (1) combination (2) text (3) audio (4) images (5) interactive
   (6) performed

2. **Answer the following questions according to the passage.**

   Open.

3. Complete each of the following statements according to the passage.

(1) combination

(2) applications

(3) advanced multimedia

(4) special effects

(5) computer-based

(6) virtual reality

4. Translate the following sentences into English.

(1) And then I do a combination of them.

(2) But not limited to that, it is still have the deeper implication.

(3) In this paper, a multimedia database engine is introduced.

(4) You must go through customs in order to pass across the border.

(5) The bridge links the island with the mainland.

(6) The small town provides a wide choice of entertainment.

(7) The school of bygone days has been replaced with that tall building.

(8) The conference included nonprofit groups, international health experts and African civic leaders.

5. Translate the following sentences into Chinese.

Omit.

Passage B

Task Ⅲ - 2

1. Match the following English phrases in Column A with their Chinese equivalents in Column B.

(1) c (2) g (3) j (4) a (5) i (6) h (7) b (8) e (9) d (10) f

2. Decide whether the following statements are T(true) or F (false) according to the passage.

(1) T (2) F (3) T (4) T (5) F (6) F (7) T (8) T (9) T (10) F

3. Translate the following passage into Chinese.

Omit.

Extra Reading

Read the following passage and choose the best answer to each question.

1. B 2. D 3. B 4. D 5. A

Section Four Writing

Task Ⅳ

1. You are required to write a certificate according to the following instructions given in Chinese.

[参考范文]

Studying Certificate

This is to certify that the student Xie Han, male, born on September 25, 1999, who came to YuCai High School, Shenyang, in September 1st, 2015 as a senior high school student. At present, he is on his second year in Senior Class 10. The personal materials will be provided by himself/herself.

YuCai High School Office of Educational Administration

Aug. 10th, 2016

2. You are required to write a farewell speech according to the information given in Chinese.

［参考范文］

Dear friends,

Good evening to everyone, and thank you all for your invitation to attend this party. I am greatly pleased to have a chance to visit your great country and meet so many of her people. This journey has been full of interesting things, and everything here impressed me a lot. I have visited factories, schools and cultural institutions. I have talked and made friends with many government officials, scientists, workers, teachers and students. My short investigation as a public servant here has given me better understanding of your country and your people.

I wish to take this opportunity to express my thanks again.

May the friendship between our two peoples be further developed!

Unit Five　E-commerce

Section One　Speaking

Open.

Section Two　Listening

Task Ⅱ

1. Listen carefully and choose the words you hear.

(1) B　(2) C　(3) A　(4) D　(5) B　(6) C　(7) D　(8) C　(9) A　(10) B

2. Listen to the short passage and choose the proper words to fill in the blanks.

(1) sports (2) front (3) leading (4) course (5) site

3. Listen to the questions and choose the best responses to what you hear.

(1) A (2) C (3) A (4) C (5) B (6) C (7) B (8) D

4. Listen to the dialogues and choose the best answer to each question you hear.

(1) B (2) C (3) C (4) A (5) D

5. Listen to the dialogue and answer the following questions by filling in the blanks.

(1) at the store (2) 270 Yuan (3) Yes, he did (4) No, it isn't

(5) the receipt

6. Listen to the passage and fill in the blanks with words you hear. Some new words given below will be of some help to you.

(1) improved (2) springing up (3) flock (4) shelves (5) provides

(6) reach (7) comfortable (8) cash (9) unnecessary (10) forget

Section Three Reading

Passage A

Task Ⅲ - 1

1. Fill in the blanks without referring to the passage.

(1) customer (2) enterprise (3) government (4) processes

(5) B2B (6) C2C

2. Answer the following questions according to the passage.

Open.

3. Complete each of the following statements according to the passage.

(1) services (2) B2B

(3) business (4) consumers

(5) C2C (6) electronic commerce users

4. Translate the following sentences into English.

(1) As is known to all, fake and inferior commodities harm the interests of consumers.

(2) We always balance the two in supply chain management.

(3) Rather than go there by air, I'd take the slowest train.

(4) We didn't place orders with this firm on account of the high price.

(5) L/C should be opened by the buyer 15 to 20 days before delivery.

(6) At the meantime, businesses should undertake social responsibilities more actively.

(7) He passed the exam with ease.

(8) You must carry out my orders.

5. Translate the following sentences into Chinese.

Omit.

Passage B

Task Ⅲ - 2

1. **Match the following English phrases in Column A with their Chinese equivalents in Column B.**

   (1) h (2) f (3) i (4) g (5) j (6) b (7) a (8) c (9) d (10) e

2. **Decide whether the following statements are T(true) or F (false) according to the passage.**

   (1) T (2) T (3) F (4) T (5) F (6) F (7) T (8) T (9) F (10) F

3. **Translate the following passage into Chinese.**

   Omit.

Extra Reading

Read the following passage and choose the best answer to each question.

1. D 2. B 3. D 4. A 5. C

Section Four Writing

Task Ⅳ

1. **You are required to write a letter according to the followinginstructions given in Chinese.**

[参考范文]

June 18th, 2017

Dear Monica,

I will be holding a party at my home on Saturday, June 25th, in order to celebrate the wedding of Mary and I.

Both of us hope, from the bottom of our hearts, that you, my best friend ever, will be able to come.

The occasion will start at 4: 00 p. m. I hope you will stay for the reception held afterward in my sitting-room. I am sure you will enjoy a good time.

We would feel honored by your presence

Yours Sincerely,

Paul

2. **You are required to write both A Letter of Invitation and A Reply to the Letter according to the following instructions given in Chinese.**

[参考范文]

- Letter 1

June 25th, 2017

Dear Prof. Smith,

I heard you would come and visit Beijing Foreign Studies University during your holiday.

Would you like to visit some famous tourist attractions? If so, I will be pleased to show you around the Great Wall, the Imperial Palace and the Summer Palace. Please let me know as soon as possible if you can come and tell me when you can make the trip.

Best wishes!

Yours,
Zhang Min

- Letter 2

June 29th, 2017

Dear Zhang Min,

Thank you for your enthusiasm and hospitality. I am very pleased to accept your invitation and plan to go to Beijing in mid-July. We can begin the trip after my 3 day's formal visit in Beijing Foreign Studies University.

Look forward with pleasure to meeting you.

Yours,
Smith

Unit Six Computer Security

Section One Speaking

Open.

Section Two Listening

Task Ⅱ

1. Listen carefully and choose the words you hear.

(1) A (2) D (3) B (4) C (5) A (6) D (7) C (8) B (9) C (10) C

2. Listen to the short passage and choose the proper words to fill in the blanks.

(1) sports (2) good (3) watching (4) baseball (5) favorite

3. Listen to the dialogues and choose the best responses to what you hear.

(1) D (2) B (3) C (4) A (5) D (6) C (7) A (8) B

4. Listen to the dialogues and choose the best answer to each question you hear.

(1) D (2) A (3) C (4) C (5) B

5. Listen to the dialogue and answer the following questions by filling in the blanks.

(1) cold (2) skating (3) No, she doesn't. (4) feel cold (5) Yes

6. Listen to the passage and fill in the blanks with words you hear. Some new words given below will be of some help to you.

(1) reasons (2) really (3) summer (4) outside (5) choice
(6) feelings (7) improve (8) suitable (9) Through (10) exercise

Section Three Reading

Passage A

Task Ⅲ - 1

1. Fill in the blanks without referring to the passage.

(1) dead (2) disconnected (3) useful (4) tradeoff (5) securing
(6) damaged

2. Answer the following questions according to the passage.

Open.

3. Complete each of the following statements according to the passage.

(1) software (2) facilities
(3) integrity (4) Data security
(5) system resources (6) tradeoff

4. Translate the following sentences into English.

(1) Listing 1 shows how this table is defined.
(2) The reasons are as follows.
(3) Unlock your phone to prevent from unwanted checks.
(4) The patient can have apples, bananas, mangos, and so forth.
(5) He's been away for at least a week.
(6) Some hours weigh against a whole lifetime, don't they?
(7) Which flat did they break into?
(8) You will stand to lose if you do business with him.

5. Translate the following sentences into Chinese.

Omit.

Passage B

Task Ⅲ - 2

1. Match the following English phrases in Column A with their Chinese equivalents in Column B.

(1) c (2) h (3) i (4) d (5) g (6) e (7) a (8) j (9) f (10) b

2. Decide whether the following statements are T(true) or F (false) according to the passage.

(1) T (2) F (3) T (4) F (5) T (6) T (7) F (8) T (9) T (10) F

3. Translate the following passage into Chinese.

Omit.

Extra Reading

Read the following passage and choose the best answer to each question.

1. B 2. A 3. C 4. C 5. A

Section Four Writing

Task Ⅳ

1. You are required to write an Inquiry Letter according to the following information given in Chinese.

[参考范文]

Mar. 29th, 2017

Dear Maria Stein,

I am interested in jet printers made by your Company. Would you please tell me further information about the prices and the after-sale services?

I'm looking forward to your early reply. Thanks a lot.

Yours,

Li Ming

2. You are required to write a Quotation Letter according to the following information given in Chinese to reply to the Inquiry Letter above.

[参考范文]

Apr. 2nd, 2017

Dear Professor Li Ming,

Thank you for your interest in our company's jet printers. A list of models and prices are attached. HP Company will offer excellent services for our clients. Besides, you can get discount if you purchase by wholesale.

Yours sincerely,

Maria Stein

Unit Seven Big Data

Section One Speaking

Open.

Section Two Listening

Task Ⅱ

1. Listen carefully and choose the words you hear.

(1) 50 (2) 140 (3) 70% (4) 200, 000, 000
(5) 13 (6) 78.21 (7) 380 (8) 70
(9) 16% (10) 5, 500, 000, 000

2. Listen to the short passage and choose the proper words to fill in the blanks.

(1) advertising (2) products (3) thousands (4) simply (5) born

3. Listen to the questions and choose the best responses to what you hear.

(1) D (2) A (3) C (4) B (5) A (6) B (7) D (8) D

4. Listen to the dialogues and choose the best answer to each question you hear.

(1) B (2) A (3) D (4) B (5) C

5. Listen to the dialogue and choose the best answer to each question you hear.

(1) D (2) B (3) C (4) C (5) D

6. Listen to the passage and fill in the blanks with words you hear. Some new words given below will be of some help to you.

(1) realize (2) count (3) evening (4) respond
(5) business (6) excitement (7) information (8) react
(9) forest (10) product

Section Three Reading

Passage A

Task Ⅲ - 1

1. Fill in the blanks without referring to the passage.

(1) data sets (2) complex (3) application (4) inadequate
(5) capturing data (6) privacy

2. Answer the following questions according to the passage.

Open.

3. Complete each of the following statements according to the passage.

(1) data sets (2) new correlations (3) information-sensingInternet of things devices
(4) big-data initiatives (5) massively parallel software (6) data management options

4. Translate the following sentences into English.

(1) How will big data change what you do?
(2) You know information and knowledge in this field update very fast.
(3) You can add more input or output columns to this data set later.
(4) Language teachers often extract examples from grammar books.
(5) Digital wealth management is a fintech category that is still climbing its way to the peak.

(6) I put it all down to his hard work and initiative.

(7) We will start out with one user database for all of our projects.

(8) You can correct any errors that you encounter in the model.

5. Translate the following sentences into Chinese.

Omit.

Passage B

Task Ⅲ - 2

1. Match the following English phrases in Column A with their Chinese equivalents in Column B.

(1) d (2) h (3) b (4) a (5) i (6) c (7) g (8) j (9) f (10) e

2. Decide whether the following statements are T(true) or F (false) according to the passage.

(1) F (2) T (3) F (4) F (5) T (6) F (7) T (8) F (9) F (10) T

3. Translate the following passage into Chinese.

Omit.

Extra Reading

Read the following passage and choose the best answer to each question.

1. D 2. A 3. C 4. D 5. B

Section Four Writing

Task Ⅳ

1. You are required to write a complaint letter according to the information given in Chinese below.

[参考范文]

July 23_{rd}, 2016

Dear Sir or Madam,

I am writing the letter to complain the camera I bought from your store last month when I was on business in Guangzhou. There I took some pictures, yet, when I had it developed after I got home, I found no pictures printed at all. I feel very frustrated about it. I have posted the camera back to you and strongly insist that you could refund me as soon as possible.

Yours sincerely,

Wang Lan

2. You are required to write an application letter according to the following instructions given in Chinese.

[参考范文]

June 25_{th}, 2016

Dear Sir or Madam,

My name is Wang Manli, 24 years old, graduating from Longjiang Technical College.

I major in Business Administration and get an excellent achievement in all my lessons. I have ever learned shorthand and typing and the speed of each is 90 words and 70 words per minute respectively. I hope to be the secretary of the general manager of your company.

Look forward to receiving your reply.

Sincerely yours,

Wang Manli

Unit Eight Cloud Computing

Section One Speaking

Open.

Section Two Listening

Task Ⅱ

1. Listen carefully and choose the words you hear.

(1) B (2) D (3) C (4) C (5) B (6) A (7) B (8) C (9) A (10) D

2. Listen to the short passage and choose the proper words to fill in the blanks.

(1) lovely (2) terrible (3) whole (4) shower (5) 13℃

3. Listen to the questions and choose the best responses to what you hear.

(1) A (2) D (3) A (4) D (5) C (6) D (7) B (8) C

4. Listen to the dialogues and choose the best answer to each question you hear.

(1) C (2) A (3) B (4) D (5) A

5. Listen to the dialogue and answer the following questions by filling in the blanks.

(1) much snow (2) Autumn (3) May (4) It is hot. (5) November

6. Listen to the passage and fill in the blanks with words you hear. Some new words given below will be of some help to you.

(1) near (2) winter (3) summer (4) long (5) sun

(6) goes down (7) warm (8) animals (9) houses (10) leave

Section Three Reading

Passage A

Task Ⅲ－1

1. Fill in the blanks without referring to the passage

(1) processing resources (2) model (3) pool (4) management effort

(5) enterprises (6) reliable

2. Answer the following questions according to the passage

Open.

3. Complete each of the following statements according to the passage.

(1) Cloud computing (2) the world (3) infrastructure costs

(4) "pay as you go" (5) utility (6) 50%

4. Translate the following sentences into English.

(1) The relationship of these devices to cloud computing may not be obvious.

(2) How do the different skill levels respond to a model like this?

(3) At this point, you have the base server set up but not your application.

(4) To go through these sections, you should have an X server up and running.

(5) This component presents a check box with configurable text to the user.

(6) Work you want to do, instead of just have to do.

(7) Last night, I signed the "pay as you go" rule into law.

(8) What kinds of virtualization should I adopt?

5. Translate the following sentences into Chinese.

Omit.

Passage B

Task Ⅲ - 2

1. Match the following English phrases in Column A with their Chinese equivalents in Column B.

(1) c (2) d (3) b (4) e (5) f (6) a (7) j (8) i (9) h (10) g

2. Decide whether the following statements are T(true) or F (false) according to the passage.

(1) T (2) F (3) T (4) F (5) F (6) T (7) T (8) F (9) F (10) T

3. Translate the following passage into Chinese.

Omit.

Extra Reading

Read the following passage and choose the best answer to each question.

1. C 2. D 3. C 4. B 5. A

Section Four Writing

Task Ⅳ

1. The following is a contribution(征稿) for an English Corner in Chinese local newspaper. Put both the introduction and the contributions into English. The title and the first sentence have been given to you.

[参考范文]

Notice

To provide with more and better English writings, we'd like to ask for contributions from readers from all walks of life. We welcome college students, English learners, foreign students and foreign friends to contribute. The writings can be:

(1) A short story;

(2) Interesting things happened around you.

Your composition will not be over 500 words.

Here is our address and e-mail:

Wang Ping, Beihai Evening Post, No. 10 Xinmin Street, Beihai

Zip Code: 230034

E-mail: wangping@163.com

2. You are required to write an Announcement according to the following information in Chinese.

[参考范文]

Announcement

A group of 35 American students will visit our school from 9:00 to 13:00 on June 22, 2016. The following are the specific arrangements to welcome the visitors.

At 8:40, we will gather to welcome the American guests at the school gate. From 9:00 to 10:30, we will hold a welcome party and exchange gifts with the American students in the reception room. Then from 10:30 to 11:30, we will show the American students our campus, library, laboratory and language lab. After that, at 11:30, we will have lunch together in the student canteen. At last, at 13:00, we will see the American students off at the school gate.

3. The following contains the main information of a notice of the Public Relations Department of a joint-venture(合资企业). You are required to write an English notice based on the following points.

[参考范文]

Notice

We are preparing to celebrate the 15th anniversary of our company. Due to the efforts and hard work of all our staff, our company is now classified as one of the first leading companies in the same industry both in scope and economic profits. To celebrate our achievement, we will hold a series of activities. Your suggestions and proposals are welcome and those being accepted will be awarded.

Please hand in your proposals to our office.

Public Relations Department

Appendix D Glossary

New Words

A

abuse [ə'bju:z] n/vt. 滥用；虐待；辱骂 (6A)
accessibility [əkˌsesə'bɪlətɪ] n. 易接近；可亲；可以得到 (8A)
accidentally [ˌæksɪ'dentəlɪ] adv. 偶然地，意外地，非故意地 (6A)
achieve [ə'tʃi:v] vt. 取得；获得；实现；成功 vi. 达到预期的目的 (8A)
actively ['æktivli] adv. 积极地；活跃地 (4A)
additional [ə'dɪʃ(ə)n(ə)l] adj. 附加的，额外的 (2B)
address [ə'dres] vt. 写姓名地址；处理 (8A)
adjacent [ə'dʒeɪs(ə)nt]adj. 邻近的，毗连的 (3B)
adjust [ə'dʒʌst] vt. 调整，使……适合；校准 vi. 调整 (8A)
administrative [əd'mɪnɪstrətɪv] adj. 管理的，行政的 (6A)
admission [əd'mɪʃn] n. 承认；准许进入；坦白；入场费 (6A)
adoption [ə'dɒpʃ(ə)n] n. 采用；收养；接受 (8A)
adult ['ædʌlt] n. 成年人 (6L)
advanced [əd'vɒ:nst] adj. 先进的；高级的 (4A)
advent ['ædvɛnt] n. 到来；出现 (5A)
advertiser ['ædvətaɪzə] n. 刊登广告者；广告客户 (7L)
advertising ['ædvətaɪzɪŋ] n. 广告；广告业 (4A)
advocate ['ædvəkeɪt] vt. 提倡，主张，拥护 n. 提倡者；支持者；律师 (8A)
aerial ['eərɪəl] adj. 空中的，航空的；空气的；空想的 n. [电讯] 天线 (7A)
afford [ə'fɔ:d] vt. 给予，提供；买得起 (6B)
air-condition [aiəkən'diʃn] n. 空调 (5L)
aisle [ail] n. 通道 (5L)
all-in-one ['ɔ: lin'wʌn] adj. 一件式的；连身的 n. 一体化 (1A)
allow [ə'laʊ] vt. 允许；给予；认可 vi. 容许；考虑 (1B)
almanacs ['ɔ:lmənæk; 'ɒl-] n. 年鉴；历书 (4A)
alternated [ɔ: l'tɜ: nətid] v. 轮流(alternate 的过去分词) adj. 间隔的，轮流的 (6A)
alternating ['ɔ:ltəneɪtɪŋ] adj. 交替的；交互的 (1A)

amendment [ə'mendmənt] n. 修正案；修改，修订（6A）
analytics [ænə'lɪtɪks] n. [化学][数] 分析学；解析学（5B）
animate ['ænɪmeɪt] vt. 使有生气；使活泼；鼓舞 adj. 有生命的（4B）
animation [ænɪ'meɪʃ(ə)n] n. 活泼，生气；激励；卡通片绘制（4A）
anticipate [æn'tɪsɪpeɪt] vt. 预期，期望；占先（6B）
appear [ə'pɪə] vi. 出现；显得；似乎；出庭；登场（2A）
applicant ['æplɪk(ə)nt] n. 申请人，申请者（6B）
application [ˌæplɪ'keɪʃ(ə)n] n. 应用；申请；应用程序（2A）
architecture ['ɒːkɪtektʃə] n. 建筑学；建筑风格；建筑式样；架构（8A）
arrangement [ə'reɪn(d)ʒm(ə)nt] n. 布置；整理；准备（3B）
array [ə'reɪ] n. 数组；排列，列阵；大批，一系列 vt. 排列（1B）
art-deco [ˌɒː t'deikəu] n. 装饰艺术（5B）
assembler [ə'semblə] n. 汇编程序；汇编机；装配工（2A）
assembly [ə'semblɪ] n. 装配；集会，集合（2A）
assign [ə'saɪn] vt. 分配；指派（6B）
assist [ə'sɪst] n. 帮助；助攻 vi. 参加；出席 vt. 帮助（2B）
associate [ə'səʊʃɪeɪt] vi. 交往；结交 n. 同事，伙伴 vt. 联想（2B）
associated [ə'soʃɪetɪd] adj. 关联的；联合的（4A）
attach [ə'tætʃ] vt. 使依附；贴上；系上；使依恋 vi. 附加（2B）
attention [ə'tenʃ(ə)n] n. 注意力；关心（5B）
attraction [ə'trækʃən] n. 吸引；吸引力（7L）
attractive [ə'træktɪv]adj. 吸引人的（5B）
auction ['ɔːkʃ(ə)n] vt. 拍卖；竞卖 n. 拍卖（5A）
audio ['ɔːdɪəʊ] adj. 声音的（4A）
audit ['ɔːdɪt] vi. 审计 n. 审计（6B）
authority [ɔː'θɒrɪtɪ] n. 权威；权力；当局（7B）
autonomic [ˌɔːtə'nɒmɪk] adj. 自律的；自治的（8A）
availability [əˌveɪlə'bɪlətɪ] n. 可用性；有效性；实用性（8A）
available [ə'veɪləb(ə)l] adj. 可获得的；可找到的；有空的（3A）

B

backplane ['bækˌ plein] n. 背板，[电子] 底板；基架（1B）
backups ['bækʌps] n. [计] 备份（6B）
barrier ['bærɪə] n. 障碍物，屏障；界线 vt. 把……关入栅栏（8B）
baseball ['beɪsbɔːl] n. 棒球（6L）
battery ['bætri] n. [电] 电池（1A）
battlefield ['bæt(ə)lfiːld] n. 战场（4A）
bench [bentʃ] n. 长凳；工作台（5L）
bent [bent] n. 爱好，嗜好（5L）

best-known ['best'nəun] adj. 流传久远的，最有名的 (3A)
beyond [bɪ'jɒnd] prep. 超过；越过；在……较远的一边 adv. 在远处 (2B)
binary ['baɪnərɪ] adj. [数] 二进制的；二元的，二态的 (2A)
biology [baɪ'ɒlədʒɪ] n. 生物；生物学 (7A)
boost [bu:st] v. 促进；增加；宣扬；推动 (5B)
broad [brɔ:d] adj. 宽的，辽阔的 n. 宽阔部分 adv. 宽阔地 (2B)
browse [braʊz] vt. 浏览 (5A)
built-in [ˌbɪlt 'ɪn] adj. 嵌入的；固定的 (1A)
bumper ['bʌmpə] adj. 丰盛的 (5L)
bundle ['bʌnd(ə)l] n. 束；捆 vt. 捆 vi. 匆忙离开 (2B)
bus [bʌs] n. (计算机) 总线 (3B)
bush [bʊʃ] n. 灌木；矮树丛 (5B)
business ['biznis] n. 企业；行业；营业；商业 (7L)
byproduct ['baɪˌprɒdʌkt] n. 副产品 (7B)

C

cable ['keɪb(ə)l] n. 缆绳；电缆；海底电报 (3A)
capable ['keɪpəb(ə)l] adj. 能干的，能胜任的 (1A)
capital ['kæpɪt(ə)l] n. 首都，省会；资金 (8B)
cash [kæʃ] n. 现金 (2L)
catastrophe [kə'tæstrəfɪ] n. 大灾难；大祸；惨败 (6B)
category ['kætɪg(ə)rɪ] n. 种类，分类；[数] 范畴 (2B)
cause [kɔ:z] n. 原因；事业；目标 vt. 引起 (2A)
celebrate ['selɪbreɪt] v. 庆祝 (1L)
central ['sentr(ə)l] adj. 中心的；主要的 n. 电话总机 (3B)
centralization [ˌsentrəlɪ'zeʃən] n. 集中化；中央集权管理 (8B)
challenge ['tʃælɪn(d)ʒ] n. 挑战；怀疑 vt. 向……挑战 (7A)
change [tʃeɪn(d)ʒ] vt. 改变；交换 n. 变化 (2A)
characteristic [kærəktə'rɪstɪk] adj. 典型的；特有的 n. 特征；特性 (8B)
charge [tʃɒ:dʒ] n. 费用；电荷 (8A)
chassis ['ʃæsɪ] n. 底盘，底架 (1B)
checkout ['tʃekaʊt] n. 检验；签出；结账台 (5A)
cherish ['tʃerɪʃ] v. 珍惜 (1L)
choice [tʃɒɪs] n. 选择；选择权；精选品 (5B)
choppy ['tʃɔpɪ] adj. 不连贯的 (3L)
chopstick ['tʃɔpstik] n. 筷子 (4L)
circuit ['sɜ:kɪt] n. 电路；线路；环形；回路 v. 巡回 (1B)
circuitry ['sɜ:kɪtrɪ] n. 电路；电路系统 (1A)
claim [kleɪm] vi. 提出要求 vt. 要求；声称；需要；认领 n. 要求 (8A)

clamshell ['klæmʃel] n. 掀盖式（1A）
classic ['klæsɪk] adj. 经典的；古典的（5B）
classifieds ['klæsifaidz] n. 分类广告（5A）
climb [klaɪm] v. 爬（1L）
code [kəʊd] n. 代码，密码；编码；法典 vt. 编码（2A）
coherence [kə(ʊ)'hɪər(ə)ns] n. 一致；连贯性；凝聚（8A）
collection [kə'lekʃ(ə)n] n. 集合（1A）
combination [kɒmbɪ'neɪʃ(ə)n] n. 结合；组合[化学]化合（1A）
command [kə'mɒ:nd] vi. 命令，指挥 vt. 命令，控制 n. 控制；命令（1A）
commercial [kə'mɜ:ʃ(ə)l] adj. 商业的（4A）
common ['kɒmən] adj. 共同的；普通的；一般的；通常的 n. 普通（1A）
commonly ['kɒmənlɪ] adv. 一般地；通常地（5B）
commonplace ['kɒmənpleɪs] n. 老生常谈；司空见惯的事 adj. 平凡的（1A）
communicate [kə'mju: nkeɪt] vt. 传达，传送（3L）
compact [kəm'pækt] n. 合同 adj. 紧凑的，紧密的；简洁的 vt. 使简洁（1A）
competition [ˌkɒmpə'tɪʃn] n. 竞争；比赛（5B）
competitor [kəm'petɪtə] n. 竞争者，对手（5B）
compiler [kəm'paɪlə] n. 编译器；[计] 编译程序；编辑者，汇编者（2A）
complementary [kɒmplɪ'ment(ə)rɪ] adj. 补足的，补充的（7B）
complexity [kəm'pleksətɪ] n. 复杂，复杂性；复杂错综的事物（6A）
component [kəm'pəʊnənt] n. 部件，组件（1A）
compromised ['kɒmprəmaɪzd] adj. 妥协的；妥协让步的，缺乏抵抗力的（6A）
computing [kəm'pju:tɪŋ] n. 计算；处理；从事电脑工作（1B）
concept ['kɒnsept] n. 观念，概念（3A）
configurable [kən'fɪgjərəbl] adj. 可配置的；结构的（8A）
configuration [kənˌfɪgə'reɪʃ(ə)n] n. 配置；结构；外形（3B）
connection [kə'nekʃn] n. 连接；关系；人脉；连接件（3A）
connectomics ['kənektɒmɪks] n. 关系学（7A）
consensual [kən'sensjʊəl] adj. 交感的；在双方同意下成立的（7B）
consideration [kənsɪdə'reɪʃ(ə)n] n. 考虑；原因；关心；报酬（7A）
consistent [kən'sɪst(ə)nt] adj. 始终如一的，一致的；坚持的（8B）
console [kən'səul] v. 安慰；抚慰；慰藉（3L）
constantly ['kɒnst(ə)ntlɪ] adv. 不断地；时常地（5B）
constitute ['kɒnstɪtju:t] vt. 组成，构成（1A）
construct [kən'strʌkt] vt. 建造，构造；创立，搭建（8B）
consultant [kən'sʌlt(ə)nt] n. 顾问；咨询者（5B）
contemporary [kən'temp(ə)r(ər)ɪ] adj. 当代的（5B）
contender [kən'tendə] n. 竞争者；争夺者（4B）
continuity [ˌkɒntɪ'nju:ɪtɪ] n. 连续性；一连串（8B）

contrast [ˈkɒntrɒːst] vi. 对比；形成对照 vt. 使对比 n. 对比（1A）
control [kənˈtrəʊl] n. 控制；管理；抑制（1B）
controlled [kənˈtrəʊld] adj. 受控制的；受约束的；克制的（6B）
controller [kənˈtrəʊlə] n. 控制器；管理员（1B）
conventional [kənˈvenʃ(ə)n(ə)l] adj. 符合习俗的，传统的；常见的（4B）
conversion [kənˈvɜːʃ(ə)n] n. 转换；变换（5B）
convert [kənˈvɜːt] vt. 使转变；转换（1A）
cooperative [kəʊˈɒpərətɪv] adj. 合作的（6B）
core [kɔː] n. 核心；要点；果心；[计] 磁心（1B）
corporation [kɔːpəˈreɪʃ(ə)n] n. 公司；法人(团体)(6B)
correlation [ˌkɒrəˈleʃən] n. 统计；相互关系（7A）
correspondence [kɒrɪˈspɒnd(ə)ns] n. 通信；一致；相当（2A）
corresponding [ˌkɒrɪˈspɒndɪŋ] adj. 相当的；一致的；通信的 v. 类似（2B）
count [kaunt] v. 计数；计算（7L）
coupled [ˈkʌpld] adj. 耦合的；连接的；成对的（8B）
crime [kraɪm] n. 罪行，犯罪；罪恶；犯罪活动 vt. 控告……违反纪律（2B）
crucial [ˈkruːʃ(ə)l] adj. 重要的；决定性的（5B）
cuddle [ˈkʌdl] v. 拥抱；搂抱（3L）
curate [ˈkjʊərət] v. 注意，组织（7B）
current [ˈkʌr(ə)nt] adj. 现在的 n.（电）流（1A）
currently [ˈkʌrəntlɪ] adv. 当前；一般地（1B）
customer [ˈkʌstəmə] n. 顾客（5A）

D

data [ˈdeɪtə] n. 数据(datum 的复数)；资料（1A）
database [ˈdeɪtəbeɪs] n. 数据库；资料库；信息库（6A）
datagram [ˈdetəˌgræm] n. 数据电报（3A）
decent [ˈdiːs(ə)nt] adj. 正派的；得体的；相当好的（4B）
decision [dɪˈsɪʒ(ə)n] n. 决定，决心；决议（5B）
deduct [dɪˈdʌkt] vt. 扣除，减去；演绎（4A）
default [dɪˈfɔːlt; ˈdiːfɔːlt] vi. 拖欠；不履行，系统默认值 vt. 不履行（2A）
define [dɪˈfaɪn] vt. 定义；使明确；规定（3B）
definition [defɪˈnɪʃ(ə)n] n. 定义（5L）
delineate [dɪˈlɪnɪeɪt] vt. 描绘；描写；画……的轮廓（7B）
delivery [dɪˈlɪv(ə)rɪ] n. [贸易] 交付（5A）
demand [dɪˈmɒːnd] vt. 要求；需要；查询 vi. 请求；查问 n. [经] 需求（8A）
density [ˈdensɪtɪ] n. 密度（7B）
deny [dɪˈnaɪ] vt. 否定，否认 vi. 否认；拒绝（6B）
dependency [dɪˈpend(ə)nsɪ] n. 属国；从属；从属物（7B）

description　[dɪˈskrɪpʃ(ə)n] n. 描述，描写；类型；说明书 (5B)
design　[dɪˈzaɪn] vt. 设计；计划　n. 设计；图案 (1A)
desktop　[ˈdesktɒp] n. 桌面；台式机 (1A)
destroy　[dɪˈstrɔɪ] vt. 破坏；消灭；毁坏 (6A)
detachable　[dɪˈtætʃəbl] adj. 可分开的；可拆开的 (1A)
detailed　[ˈdiːteɪld] adj. 详细的，精细的 (5B)
detect　[dɪˈtekt] v. 发现 (7B)
development　[dɪˈveləpm(ə)nt] n. 发展；开发 (6A)
deviation　[diːvɪˈeɪʃ(ə)n] n. 偏差；误差；背离 (1A)
device　[dɪˈvaɪs] n. 装置；策略；图案 (1B)
differ　[ˈdɪfə] vt. 使……相异；使……不同　vi. 相异；意见分歧 (3A)
digital　[ˈdɪdʒɪt(ə)l] adj. 数字的；手指的　n. 数字；键 (1B)
direct　[dəˈrekt] adj. 直接的；恰好的　vt. 导演；指向　vi. 指导；指挥 (1A)
direction　[dɪˈrekʃn, daɪ-;] n. 方向；指导；趋势；用法说明 (3B)
disaster　[dɪˈzɒːstə(r)] n. 灾难；彻底的失败；不幸；祸患 (6B)
disc　[dɪsk] n. 圆盘，[电子] 唱片(等于 disk) (1B)
disconnected　[dɪskəˈnektɪd] adj. 分离的；不连贯的；无系统的 (6A)
display　[dɪˈspleɪ] n. 显示；炫耀　vt. 显示；表现；陈列 (1B)
disrupt　[dɪsˈrʌpt] vt. 破坏；使瓦解；使分裂；使中断；使陷于混乱 (2B)
dissect　[daɪˈsekt; dɪ-] vt. 切细；仔细分析　vi. 进行解剖 (4A)
distance　[ˈdɪstəns] n. 距离；远方；疏远；间隔　vt. 疏远 (8A)
distinguish　[dɪˈstɪŋgwɪʃ] v. 区分；辨别 (5B)
distribute　[dɪˈstrɪbjuːt] vt. 分配；散布；分开 (8B)
diverse [daɪˈvɜːs; ˈdaɪvɜːs] adj. 不同的；多种多样的；变化多的 (7B)
divide　[dɪˈvaɪd] v. 分成 (2L)
documentation　[ˌdɒkjʊmenˈteɪʃ(ə)n] n. 文件，证明文件，文件编制 (2A)
domain　[də(ʊ)ˈmeɪn] n. 领域；域名；产业；地产 (2B)
dominant　[ˈdɒmɪnənt] adj. 显性的；占优势的；支配的，统治的　n. 显性 (2A)
dream　[driːm] v. /n. 梦想；梦 (1L)
drugstore　[ˈdrʌgstɔː] n. [药] 药房 (4L)
duplicate　[ˈdjuːplɪkeɪt] v. 复制　n. 副本；复制品　adj. 复制的 (5B)
during　[ˈdjuəriŋ] prep. (表示时间)在……期间 (4L)
dynamic　[daɪˈnæmɪk] adj. 动态的；动力的；有活力的　n. 动力 (8B)

E

E-commerce　[iːkɒmərs] n. 电子商务 (5A)
economy　[ɪˈkɒnəmɪ] n. 经济；节约；理财 (8A)
ecosystem　[ˈiːkəʊsɪstəm] n. 生态系统 (7A)
edit　[ˈedɪt] vt. 编辑；校订　n. 编辑工作 (4B)

effect [ɪ'fekt] n. 影响；效果；作用 vt. 产生；达到目的（2A）
efficiency [ɪ'fɪʃ(ə)nsɪ] n. 效率；效能（5A）
efficient [ɪ'fɪʃ(ə)nt] adj. 有效率的；有能力的；生效的（1B）
effort ['efət] n. 努力；成就（8A）
elapse [ɪ'læps] vi. 消逝；时间过去 n. 流逝（7B）
elasticity [elæ'stɪsɪtɪ] n. 弹性；弹力；灵活性（8B）
electronically [ˌɪlek'trɒnɪklɪ] adv. 电子地（5A）
element ['elɪm(ə)nt] n. 元素；要素；原理（3B）
embosser [im'bɔsə] n. [皮革] 压纹机；[皮革] 压花机（1B）
emerge [ɪ'mɜːdʒ] vi. 浮现；摆脱；暴露（1A）
enclosure [ɪn'kləʊʒə] n. 附件；围墙；围场（1A）
encompass [ɪn'kʌmpəs] vt. 包含；包围，环绕；完成（7B）
encounter [ɪn'kaʊntə] vt. 遭遇，邂逅；遇到 n. 遭遇（7A）
encourage [ɪn'kʌrɪdʒ] vt. 鼓励，怂恿；激励；支持（6B）
encryption [ɪn'krɪpʃn] n. 加密；编密码（6B）
encyclopedia [ɪnˌsaɪklə'piːdiə] n. 百科全书（4A）
engineering [endʒɪ'nɪərɪŋ] n. 工程，工程学 v. 设计（2A）
enhance [ɪn'hɒːns] vt. 提高；加强；增加（7B）
enormous [ɪ'nɔːməs] adj. 庞大的，巨大的（3A）
enter ['entə] n. [计] 输入；回车 vt. 进入；开始；参加 vi. 进去（1B）
enterprise ['entəpraɪz] n. 企业；事业（5A）
entertainment [ˌentə'teɪnmənt] n. 娱乐；消遣；款待（2B）
enthusiasm [in'θjuiziæzəm] n. 热心，热情，热忱（5L）
entity ['entɪtɪ] n. 实体（5A）
environment [in'vaiərənmənt] n. 环境（5L）
environmental [ˌɪnvaɪrən'ment(ə)l] adj. 环境的，周围的；有关环境的（7A）
essential [ɪ'senʃ(ə)l] adj. 基本的；必要的；本质的 n. 本质；要素（2B）
establish [ɪ'stæblɪʃ; e-] vt. 建立；创办；安置（3A）
evaluate [ɪ'væljʊeɪt] v. 评价；估价（6B）
exabyte [eksəˌbaɪt] n. 百亿亿字节，艾字节（7A）
exchange [ɪks'tʃeɪndʒ; eks-] n. 交换；交流；交易所；兑换（3A）
excitement [ɪk'saɪtmənt] n. 激动；兴奋（7L）
executable [ɪg'zekjʊtəb(ə)l] adj. 可执行的；可实行的（2A）
execute ['eksɪkjuːt] vt. 实行；执行（1A）
executive [ɪg'zekjʊtɪv] adj. 经营的；执行的 n. 总经理；执行者（7A）
expand [ɪk'spænd] vt. 扩张；使膨胀；详述 vi. 发展；张开，展开（1B）
expansion [ɪk'spænʃ(ə)n] n. 膨胀（1A）
expenditure [ɪk'spendɪtʃə; ek-] n. 支出，花费；经费，消费额（8B）
exposed [ɪk'spəʊzd] adj. 暴露的 v. 暴露，揭露(expose 的过去分词)(6A)

external [ɪk'stɜ:n(ə)l; ek-] adj. 外部的；外国的 n. 外部；外面（1B）
externally [ɪk'stə:nli] adv. 外部地；外表上，外形上（1B）
extract ['ekstrækt] vt. 提取；取出；摘录；榨取 n. 汁；摘录（1B）

F

face-to-face ['feɪsˌtʊ'feɪs] adj. 面对面的 adv. 面对面地（5A）
fact [fækt] n. 事实，现实（7L）
factor ['fæktə] n. 因素；要素（1A）
fascinating ['fæsineɪtɪŋ] adj. 迷人的；吸引人的（7L）
feature ['fi:tʃə] n. 特征；容貌 vi. 起重要作用 vt. 特写；以……为特色（1B）
feedback ['fi:dbæk] n. 反馈；成果，资料；回复（5B）
file [faɪl] n. 文件；档案；文件夹 vt. 提出；琢磨；把……归档（1B）
fine-grained ['fain-'greind] adj. 细粒的 n. 细粒度（8B）
fintech ['faɪntek] n. 金融科技（7A）
firewall ['faɪɚwɔl] n. 防火墙（6B）
fit [fɪt] vt. 安装；使……适应（1A）
flash [flæʃ] n. 闪光；交光灯；瞬间场面，瞬间镜头（1B）
flexibility [ˌfleksɪ'bɪlɪtɪ] n. 灵活性；弹性；适应性（8B）
fluctuate ['flʌktʃʊeɪt] vi. 波动；涨落；动摇 vt. 使波动（8A）
fool [fu:l] n. 傻瓜（5L）
fork [fɔ: k] n. 叉（4L）
form [fɔ:m] n. 形式，形状 vt. 构成，组成；排列（1A）
format ['fɔ: mæt] n. [计] 格式（4A）
forum ['fɔ:rəm] n. 论坛，讨论会（5A）
freshly ['freʃli] adv. 气味清新地（4L）
friendship ['frendʃɪp] n. 友谊（1L）
function ['fʌŋ(k)ʃ(ə)n] n. 功能；[数] 函数；职责 vi. 运行（1B）
functionality [fʌŋkʃə'næləti] n. 功能；[数] 泛函性，函数性（1B）
fundamental [fʌndə'ment(ə)l] adj. 基本的，根本的 n. 基本原理（1B）
fusion ['fju:ʒ(ə)n] n. 融合；熔化；熔接；融合物（7B）

G

gain [gein] v. 获得；吸引（7L）
garbage ['gɒ:bɪdʒ] n. 垃圾；废物（6A）
generally ['dʒen(ə)rəlɪ] adv. 通常；普遍地，一般地（1A）
generate ['dʒenəreɪt] vt. 使形成；发生（7A）
generation [dʒenə'reɪʃ(ə)n] n. 一代；产生（3L）
genomics [dʒə'nɒmɪks] n. 基因组学；基因体学（7A）
geometric [ˌdʒɪə'metrɪk] adj. 几何学的；[数] 几何学图形的（3B）

geometry [dʒɪ'ɒmɪtrɪ] n. 几何学，几何结构 (3B)
gigabyte ['gɪgəbaɪt] n. 十亿字节；十亿位组 (1B)
goal [gəʊl] n. 目标；球门，得分数；终点 (2B)
government ['gʌvənmənt] n. 政府 (5A)
governmental [ˌgʌvən'mɛntl] adj. 政府的；政治的 (4A)
grab [græb] v. 攫取；霸占 (4A)
graduate ['grædjueɪt] v. 毕业 (1L)
graduation [ˌgrædju'eɪʃən] n. 毕业 (1L)
graphic ['græfɪk] adj. 形象的；图表的；绘画似的 (1A)
grid [grɪd] n. 网格；格子，栅格；输电网 (8A)
group [gru:p] n. 组；分(类) adj. 群的；团体的 vi. 聚合 把……分组 (2B)

H

habit ['hæbit] v. 习惯 (4L)
hamburger ['hæmbə:gə] n. 汉堡 (4L)
handcart ['hændkɒ:t] n. 手推车 (5L)
hardware ['hɒ:dweə] n. 计算机硬件 (1A)
harm [hɒ:m] n. 伤害；损害 vt. 伤害 (2B)
healthy ['helθɪ] adj. 健康的 (6L)
high-level [ˌhaɪ'levl] adj. 高级的；高阶层的；在高空的 (2A)
hindrance ['hɪndr(ə)ns] n. 障碍；妨碍 (6B)
host [həʊst] n. [计] 主机 (3A)
house [haʊs] n. 住宅；家庭；机构 vt. 覆盖；给……房子住 (1A)
human ['hju:mən] adj. 人的；人类的 n. 人；人类 (1B)

I

identification [aɪˌdentɪfɪ'keɪʃ(ə)n] n. 鉴定，识别；认同；身份证明 (6B)
illustration [ɪlə'streɪʃ(ə)n] n. 说明；插图；例证；图解 (3B)
iMac abbr. Integrated Microwave Amplifier Converter；苹果一体机 (1A)
image ['ɪmɪdʒ] n. 影像；想象 (1B)
immerse [ɪ'mə:s] vt. 沉浸；使陷入 (4A)
improved [ɪm'prʊvd] adj. 改良的；改进过的 (5A)
inadequate [ɪn'ædɪkwət] adj. 不充分的，不适当的 (7A)
incorporate [ɪn'kɔ:pəreɪt] vt. 包含，吸收；体现；把……合并 (4A)
indirectly [ˌɪndɪ'rek(t)lɪ] adv. 间接地；不诚实；迂回地 (3B)
individual [ˌɪndɪ'vɪdʒʊəl] adj. 个人的；个别的；独特的 n. 个人，个体 (2A)
induce [in'djuis] v. 引诱，劝 (5L)
inductive [ɪn'dʌktɪv] adj. [数] 归纳的；[电] 感应的；诱导的 (7B)
infancy ['ɪnf(ə)nsɪ] n. 初期；婴儿期；幼年 (8A)

influenced [ˈɪnflʊənsd] adj. 受影响的 (5B)
informatics [ˌɪnfəˈmætɪks] n. [计] 信息学；情报学(复数用作单数) (7A)
information [ɪnfəˈmeɪʃ(ə)n] n. 信息，资料；知识；情报；通知 (1B)
infrastructure [ˈɪnfrəstrʌktʃə] n. 基础设施；公共建设；下部构造 (8A)
infrequent [ɪnˈfriːkw(ə)nt] adj. 罕见的；稀少的；珍贵的；不频发的 (8B)
initiative [ɪˈnɪʃɪətɪv] n. 主动权 adj. 主动的；自发的；起始的 (7A)
innovative [ˈɪnəvətɪv] adj. 革新的，创新的；新颖的 (7B)
insert [ɪnˈsɜːt] vt. 插入；嵌入 n. 插入物；管芯 (1B)
insight [ˈɪnsaɪt] n. 洞察力；洞悉 (7B)
instant [ˈɪnst(ə)nt] adj. 立即的；紧急的 n. 瞬间；立即；片刻 (3A)
instruction [ɪnˈstrʌkʃənz] n. 指令；说明 (1A)
integrate [ˈɪntɪgreɪt] vt. 使……完整；使……成整体 (4A)
integrated [ˈɪntɪgreɪtɪd] adj. 综合的；完整的；互相协调的 (1A)
integration [ɪntɪˈgreɪʃ(ə)n] n. 集成；综合 (7B)
integrity [ɪnˈtegrəti] n. 完整；正直，诚实；[计算机] 保存；健全 (6A)
intensive [ɪnˈtensɪv] adj. 加强的；集中的；透彻的；加强语气的 (8B)
interaction [ɪntərˈækʃ(ə)n] n. 相互作用 (4A)
interactive [ɪntərˈæktɪv] adj. 交互式的 (4A)
intercept [ˌɪntəˈsept] v. 拦截 (6B)
interface [ˈɪntəfeɪs] n. 界面；接口；接触面(8B)
intermediate [ˌɪntəˈmiːdɪət] adj. 中间的，中级的 (3B)
internal [ɪnˈtɜːn(ə)l] n. 本质 adj. 内部的 (1A)
internal [ɪnˈtɜːn(ə)l] n. 本质 adj. 内部的；体内的；(机构)内部的 (1B)
international [ɪntəˈnæʃ(ə)n(ə)l] adj. 国际的 (3L)
interpret [ɪnˈtɜːprɪt] vt. 说明；口译 vi. 解释；翻译 (3A)
interpreter [ɪnˈtɜːprɪtə] n. 解释者；口译者；注释器 (2A)
interrupt [ɪntəˈrʌpt] vt. 中断；打断；插嘴；妨碍 (2A)
involve [ɪnˈvɒlv] vt. 包含；牵涉；使陷于 (5A)
irreplaceable [ɪrɪˈpleɪsəb(ə)l] adj. 不能替代的，不能调换的 (6B)

J

joystick [ˈdʒɔɪˌstɪk] n. (计算机游戏的)操纵杆，摇杆，控制杆 (1B)

K

key [kiː] n. (打字机等的)键；关键；钥匙 vt. 键入 (8B)
keyboard [ˈkiːbɔːd] n. 键盘 (1A)

L

LAN [læn] n. 局域网 (6B)

language [ˈlæŋgwɪdʒ] n. 语言；语言文字；表达能力 (2A)
laptop [ˈlæptɒp] n. 膝上型轻便电脑，笔记本电脑 (1A)
laurel [ˈlɒr(ə)l] n. 月桂属植物，月桂 (5B)
lax [læks] adj. 松的；松懈的；腹泻的 (6B)
layer [ˈleɪə] n. 层，层次； vt. 把……分层堆放 (3A)
layout [ˈleɪaʊt] n. 布局；设计；安排；陈列 (3B)
leather [ˈleðə] n. 皮革；皮革制品 (5L)
legally [ˈliːgəlɪ] adv. 法律上，合法地 (6A)
library [ˈlaɪbrərɪ] n. 图书馆，藏书室；文库(2A)
limitation [lɪmɪˈteɪʃ(ə)n] n. 限制；限度；极限 (7A)
literature [ˈlɪtərɪtʃə] n. 文学；文学作品 (7L)
litter [ˈlɪtə] n. 垃圾 (6L)
location [lə(ʊ)ˈkeɪʃ(ə)n] n. 位置；地点 (2A)
logical [ˈlɒdʒɪk(ə)l]adj. 合逻辑的，合理的；逻辑学的 (3B)
loop [luːp] n. 环；圈 v. 使成环；以环连接 (3B)
loosely [ˈluːslɪ] adv. 宽松地；放荡地；轻率地 (8B)
loss [lɒs] n. 减少；亏损；失败；遗失 (6A)
lounge [laʊn(d)ʒ] n. 休息室；闲逛；躺椅 (5B)
lower [ˈləʊə] vt. 减弱，减少，降下；贬低 vi. 降低；减弱；跌落 (8B)
low-pitched [ˈləupɪtʃt] adj. (声音)低沉的 (3L)
low-voltage [ˈləuˈvəultidʒ] adj. 低压的 (1A)

M

machine [məˈʃiːn] n. 机械，机器；机械般工作的人 vt. 用机器制造 (1B)
macro [ˈmækrəʊ] adj. 巨大的，大量的 n. 宏，巨(计算机术语)(5A)
maintain [meinˈtein] vt. 维持；维修；供养 (6B)
maintenance [ˈmeɪntənəns] n. 维护，维修；保持 (2B)
majority [məˈdʒɒrɪtɪ] n. 多数；成年 (1B)
maliciously [məˈlɪʃəslɪ] adv. 有敌意地 (6A)
malware [mælwɛə] n. 恶意软件 (2B)
manage [ˈmænɪdʒ] vt. 管理；经营；控制；设法 vi. 处理 (2B)
manageability [ˌmænɪdʒəˈbɪləti] n. 易处理；易办；顺从 (8A)
management [ˈmænɪdʒm(ə)nt] n. 管理；管理人员；管理部门 (5A)
massively [ˈmæsivli] adv. 大量地；沉重地；庄严地 (7A)
mathematics [ˌmæθəˈmætɪks] n. 数学 (1L)
maturity [məˈtʃʊərətɪ] n. 成熟；到期；完备 (7B)
mechanism [ˈmek(ə)nɪz(ə)m] n. 机制；原理，途径 (5A)
media [ˈmiːdɪə] n. 媒体；媒质(medium 的复数) (1B)
medium [ˈmiːdɪəm] adj. 中等的；半生熟的 n. 方法；媒体；媒介；(3A)

meld [meld] vi. 合并；混合 vt. 使……合并；使……混合 n. 结合；融合 (4A)
mesh [meʃ] n. 网眼；网丝；圈套 (3B)
metal ['met(ə)l] n. 金属；合金 vt. 以金属覆盖 adj. 金属制的 (1A)
meteorology [ˌmiːtɪə'rɒlədʒɪ] n. 气象状态，气象学 (7A)
microphone ['maɪkrəfəʊn] n. 扩音器，麦克风 (1B)
minimum ['mɪnɪməm] n. 最小值；最低限度；最小量 adj. 最小的 (2B)
misuse [mɪs'juːz] vt. /n. 滥用；误用；虐待 (6A)
mobile ['məʊbaɪl] adj. 可移动的；机动的 (3L)
model ['mɒdl] n. 模型；典型；模特儿；样式 (5A)
modify ['mɒdɪfaɪ] v. 修改，修饰；更改 (6A)
monitor ['mɒnɪtə] n. 监视器；显示屏；班长 (1A)
motherboard ['mʌðɚbɔrd] n. 底板；母板；主板 (1A)
mountain ['mauntɪn] n. 山 (1L)
multi-core ['mʌtɪkɒ] n. 多核，多核心 (2A)
multiple ['mʌltɪpl] adj. 多重的；多样的；许多的 (8B)
multitenancy [ˌmʌtlɪ'tenəsɪ] n. 多租户技术 (8B)

N

natural ['nætʃrəl] adj. 自然的；物质的；天生的 n. 自然的事情 (2A)
naturally ['nætʃərəli] adv. 自然地 (5L)
nature ['neɪtʃə] n. 自然；性质；本性；种类 (3B)
navigation [nævɪ'geɪʃ(ə)n] n. 航行；航海 (4B)
neatly ['niːtlɪ] adj. 整洁地，干净地 (2L)
network ['netwɜːk] n. 网络；广播网；网状物 (3A)
networked ['netwəːkt] adj. 网路的；广播电视联播的 (3A)
node [nəʊd] n. 节点 (3A)
non-executable [ˌnʌnɪg'zekjʊtəb(ə)l] adj. 不可执行的；不可实行的 (2A)
nonlinear [nɒn'lɪnɪə] adj. 非线性的 (7B)
nonprofit [nɒn'prɒfɪt] adj. 非营利的 (4A)
noodle ['nuːdl] n. 面条 (4L)
normally ['nɔːm(ə)lɪ] adv. 正常地；通常地 (1A)
numerous ['njuːm(ə)rəs] adj. 许多的，很多的 (5B)

O

object ['ɒbdʒɪkt] n. 目标；物体；宾语 vi. 反对 (1A)
objective [əb'dʒektɪv] adj. 客观的；目标的 n. 目的；目标 (5B)
observable [əb'zɜːvəbl] adj. 觉察得到的；看得见的 n. 感觉到的事物 (2A)
observe [əb'zɜːv] vt. 庆祝 vt. 观察；遵守 (7B)
obtain [əb'teɪn] vi. 获得；流行 vt. 获得 (1B)

offer ['ɒfə] vt. 提供；试图 n. 提议；出价；录取通知书 vi. 出现（1B）
one-time ['wʌntaɪm] adj. 以前的；一次性的 adv. 一度；从前（8B）
online [ɒn'laɪn] adj. 联机的；在线的 adv. 在线地（2A）
operation [ɒpə'reɪʃ(ə)n] n. 操作；经营；[外科] 手术；[计] 运算（1B）
optical ['ɒptɪk(ə)l] adj. 光学的；眼睛的，视觉的（1B）
optimize ['ɒptɪmaɪz] vt. 使最优化，使完善 vi. 优化（5B）
option ['ɒpʃ(ə)n] n. [计] 选项；选择权；买卖的特权（5A）
order ['ɔ：də] v. 点菜，预订（2L）
organic [ɔ:'gænɪk] adj. 组织的；器官的；根本的（5B）
organization [ˌɔ:gənaɪ'zeɪʃn] n. 组织；机构；体制；团体（8A）
organizational [ˌɔ:gənaɪ'zeɪʃənl] adj. 组织的；编制的（3A）
outcome ['aʊtkʌm] n. 结果，结局；成果（7B）
overload [əuvə'ləud] v. 使超载；使负荷过重（3L）
overwhelm [ˌəuvə'welm] v. 压倒；击败；征服（3L）

P

paint [peɪnt] n. 油漆；颜料，涂料（5A）
parallel ['pærəlel] n. 平行线；对比 vt. 使……与……平行 adj. 平行的（7A）
partial ['pɒ:ʃ(ə)l] adj. 局部的；偏爱的；不公平的（3B）
participant [pɒ:'tɪsɪp(ə)nt] n. 参与者；关系者（5A）
participate [pɒ:'tɪsɪpeɪt] vi. 参与，参加；分 vt. 分享；分担（4A）
particular [pə'tɪkjʊlə(r)] adj. 特别的；详细的；独有的 n. 详细说明（2A）
parts [pɒ:ts] n. [机] 零件，部件；形式（1A）
party ['pɒ:tɪ] n. 政党，党派；聚会，派对；当事人（5A）
passive ['pæsɪv] adj. 被动的，消极的（4A）
password ['pɒ:swɜ:d] n. 密码；口令（6B）
pastime ['pɒ:staɪm] n. 娱乐；消遣（4A）
path [pæθ] n. 道路；小路；轨道（3B）
payment ['peɪm(ə)nt] n. 付款，支付（5A）
penalize ['pi：nəlaɪz] vt. 处罚；处刑；使不利（5B）
per-capita [pə 'kæpɪtə] n. [统计] 人均；（拉丁）每人；按人口计算（7A）
perform [pə'fɔ:m] vt. 执行；完成；演奏 vi. 执行，机器运转（2B）
performance [pə'fɔ:m(ə)ns] n. 性能；绩效；表演；执行；表现（1A）
peripheral [pə'rɪf(ə)r(ə)l] adj. 外围的；次要的 n. 外部设备（1A）
permanently ['pɜ:m(ə)nəntlɪ] adv. 永久地，长期不变地（1B）
personal ['pɜ:s(ə)n(ə)l] adj. 个人的；亲自的 n. 人称代名词（1A）
personnel [ˌpɜ：sə'nel] n. 全体员工；人员；人事部门（6A）
physical ['fɪzɪk(ə)l] adj. [物] 物理的；身体的；物质的（1A）
physically ['fɪzɪkəllɪ] adv. 身体上，身体上地（3B）

picnic [ˈpɪknɪk] n. 野餐 (1L)
pile [paɪl] n. 堆 (2L)
pitfall [ˈpɪtfɔːl] n. 陷阱，圈套；缺陷 (8A)
plastic [ˈplæstɪk] adj. 塑料的 n. 塑料制品 (1A)
platform [ˈplætfɔːm] n. 平台；月台 (4A)
plum [plʌm] n. 李子 (5L)
politely [pəˈlaɪtlɪ] adj. 有礼貌地 (2L)
pool [puːl] n. 联营；撞球；共同资金 vi. 联营，合伙经营 (8A)
port [pɔːt] n. (计算机的)端口 (1A)
porting [pɔrtɪŋ] n. 移植，出口商品 v. 携带；搬运(port 的现在分词) (1B)
possibility [ˌpɒsɪˈbɪlɪtɪ] n. 可能性；可能发生的事物 (5B)
post- [pəust-] pref. 表示“后、在……之后”的意思 (1B)
potential [pəˈtenʃl] n. 潜能；可能性 adj. 潜在的 (5B)
potentially [pəˈtɛnʃəli] adv. 可能地，潜在地 (2B)
PowerPoint [ˈpauəpɔint] n. 微软办公软件 (4B)
practice [ˈpræktɪs] n. 实践；练习；惯例 vi. 练习 (2B)
practitioner [prækˈtɪʃ(ə)nə] n. 开业者，从业者 (7A)
preceding [prɪˈsiːdɪŋ] adj. 在前的；前述的 v. 在……之前 (2A)
predictive [prɪˈdɪktɪv] adj. 预言性的；成为前兆的 (7A)
presentation [prez(ə)nˈteɪʃ(ə)n] n. 展示；描述 (4A)
prevent [prɪˈvent] v. 防止；预防 (7L)
printer [ˈprɪntə] n. [计] 打印机；印刷工 (1B)
procedure [prəˈsiːdʒə] n. 程序，手续；步骤 (6B)
process [prəˈses] vt. 处理；加工 (2A)
processor [ˈprəusesə] n. [计] 处理器；处理程序；加工者 (2A)
productivity [prɒdʌkˈtɪvɪtɪ] n. 生产力；生产率；生产能力 (8B)
professional [prəˈfeʃ(ə)n(ə)l] adj. 专业的；职业的 n. 专业人员 (5B)
program [ˈprəugræm] n. 程序；计划 vt. 为……制订计划 vi. 编程序 (2A)
programmer [ˈprəugræmə] n. (计) 程序设计员 (2A)
programming [ˈprəugræmɪŋ] n. 设计，规划[计] 程序编制 (4B)
promoter [prəˈməutə] n. 促进者；发起人 (3L)
proponent [prəˈpəunənt] n. 支持者；建议者；提出认证遗嘱者 (8A)
protocol [ˈprəutəkɒl] n. 协议；草案；礼仪 (3A)
provide [prəˈvaɪd] vt. 提供；规定；准备；装备 vi. 规定；抚养 (2A)
provision [prəˈvɪʒ(ə)n] n. 规定；准备 vt. 供给……食物及必需品 (8A)
provisioning [prəˈvɪʒən] n. 准备金提取 v. 供应补给品 (8B)
purchase [ˈpɜːtʃəs] n./v. 购买 (5A)
purchase [ˈpɜːtʃəs] n. 购买；紧握 vt. 购买；赢得(5B)
purportedly [ˈpɜːpətɪdli] adv. 据称，据称地 (8B)

purpose [ˈpɜ:pəs] n. 目的；用途；意志 vt. 决心（2B）

Q

quality [ˈkwɒlətɪ] n. 质量，[统计] 品质；特性（5B）
query [ˈkwɪərɪ] n. 疑问，质问；疑问号；[计] 查询 v. 询问（5B）

R

range [reɪn(d)ʒ] n. 范围；幅度；排；山脉 vi. 平行；延伸 vt. 归类于（2B）
rapidly [ˈræpɪdlɪ] adv. 迅速地；很快地；立即（8A）
rating [ˈreɪtɪŋ] n. 等级；等级评定（5B）
raw [rɔ:] adj. 生的；未加工的（5A）
react [rɪˈækt] v. 起反应；起作用（7L）
readable [ˈri:dəb(ə)l] adj. 可读的；易读的（1B）
realistic [rɪəˈlɪstɪk] adj. 现实的（4A）
realistically [ˌriəˈlɪstɪkli] adv. 现实地；实际地（2A）
real-time [ˌrɪəl ˈtaɪm] adj. 实时的；接到指示立即执行的（8B）
rearrange [ri:əˈreɪn(d)ʒ] vt. 重新排列；重新整理（4L）
recipient [rɪˈsɪpɪənt] n. 收件人（4A）
recognition [ˌrekəgˈnɪʃn] n. 认识，识别；认可；褒奖；酬劳（6A）
recovery [rɪˈkʌv(ə)rɪ] n. 恢复，复原（6B）
reduced [rɪˈdju:st] adj. 减少的（5A）
redundant [rɪˈdʌnd(ə)nt] adj. 多余的，过剩的；冗长的，累赘的（8B）
reference [ˈref(ə)r(ə)ns] n. 参考，参照（4A）
refuse [rɪˈfju: z] v. 拒绝（2L）
register [ˈredʒɪstə] v. 登记；注册（2L）
regression [rɪˈgreʃ(ə)n] n. 回归；退化；逆行；复原（7B）
relate [rɪˈleɪt] vt. 叙述；使……有联系 vi. 涉及；认同；符合（2A）
relative [ˈrelətɪv] adj. /n. 相对的；相关的；比较而言的 n. 亲属（6A）
relatively [ˈrelətɪvlɪ] adv. 相当地；相对地，比较地（1A）
release [rɪˈli:s] vt. 释放；发射；让与 n. 释放；发布；让与（8A）
relevant [ˈreləvənt] adj. 相关的；切题的；中肯的（4A）
reliability [rɪˌlaɪəˈbɪlətɪ] n. 可靠性（1B）
reliable [rɪˈlaɪəb(ə)l] adj. 可靠的；可信赖的 n. 可靠的人（8A）
reliably [riˈlaiəbli] adv. 可靠地；确实地（5B）
remote [rɪˈməʊt] adj. 遥远的 n. 远程（6B）
removable [rɪˈmu:vəbl] adj. 可移动的；可去掉的；可免职的（1B）
replace [rɪˈpleɪs] vt. 取代，代替（6B）
require [rɪˈkwaɪə] vt. 需要；要求；命令（2A）
resource [rɪˈsɔ:s; rɪˈzɔ:s] n. 资源，财力；办法；智谋（2B）

respond [ris'pɔnd] v. 相应；反应 (7L)
restore [rɪ'stɔː] v. 恢复；修复 (6A)
restriction [rɪ'strɪkʃ(ə)n] n. 限制；约束；束缚 (6B)
retain [rɪ'teɪn] vt. 保持；雇；记住 (8B)
reveal [rɪ'viːl] vt. 显示；透露 (7B)
revenue ['revənjuː] n. 税收，国家的收入；收益 (5B)
review [rɪ'vjuː] n. 回顾；复习；评论 (5B)
revisionism [rɪ'vɪʒənɪzəm] n. 修正主义 (7B)
rude [ruːd] adj. 粗鲁的；不礼貌的 (4L)

S

safeguard ['seɪfgɒːd] n. 保护，保卫；防护措施 vt. 防护；保护 (6A)
salad ['sæləd] n. 沙拉 (4L)
scalability [ˌskeilə'biliti] n. 可扩展性；可伸缩性；可量测性 (8A)
scalability [ˌskeilə'biliti] n. 可扩展性；可伸缩性；可量测性 (8B)
scale [skeɪl] n. 规模；比例；刻度；天平；数值范围 vi. 衡量 (8A)
scanner ['skænə] n. (计) 扫描仪；扫描器 (1B)
scheme [skiːm] n. 计划；组合；体制；诡计 (3B)
screening ['skriːnɪŋ] n. 筛选；审查 v. 筛选 adj. 筛选的 (6B)
seasonal ['siːz(ə)n(ə)l] adj. 季节的；周期性的 (6L)
security [sɪ'kjʊərəti] n. 安全；保证；防护 (6A)
seem [siːm] v. 好像 (1L)
select [sɪ'lekt] v. 挑选；选拔 (5A)
semi-circle [sɛmɪˌsɜːkəl] n. 半圆 (5B)
semi-structured [semɪ'strʌktʃəd] adj. 半结构化的 (7B)
sequence ['siːkw(ə)ns] n. [数][计] 序列；顺序 vt. 按顺序排好 (4B)
server ['sɜːvə] n. (计算机)主机，发球员；服伺者 (1B)
shell [ʃel] n. 壳(命令解析器) (2B)
shoot [ʃuːt] vt. 射击，射中 (3L)
signal ['sɪgnl] n. 信号；动机；标志 (3A)
signify ['sgnɪfaɪ] n. 代表；预示 v. 象征；预示 (2A)
silver ['sɪlvə] n. 银；银器 (5L)
similar ['sɪmɪlə] adj. 相似的 n. 类似物 (1A)
simply ['sɪmplɪ] adv. 简单地；仅仅；简直；朴素地 (2A)
simulations [ˌsɪmjʊ'leɪʃən] n. [计] 模拟(simulation 的复数)；[计] 仿真 (7A)
simultaneously [ˌsɪml'teɪnɪəslɪ] adv. 同时地 (8B)
sip [sɪp] v. 小口喝 (2L)
skin [skɪn] n. 皮，皮肤 (8L)
slide [slaɪd] n. 滑动；幻灯片；滑梯；雪崩 v. 滑动 (4B)

slot [slɒt] n. 位置；插槽 (1A)
snippet ['snɪpɪt] n. 小片；片断 (5B)
so-called [səʊ'kɔ：ld] adj. 所谓的；号称的 (3B)
software ['sɒf(t)weə] n. 软件 (1A)
solid-state ['sɒlɪdˌsteɪt] adj. 固态的；固态电子学的；使用电晶体的 (1B)
solution [sə'lʊʃən] n. 解决方案；应对措施 (6A)
sorter ['sɔrtɚ] n. 从事分类的人；分类机 (4B)
soundtrack ['saʊn(d)træk] n. 声带；声道；声迹；电影配音 (4B)
speaker ['spi:kə] n. 扬声器；说话者；演讲者 (1B)
special ['speʃ(ə)l] n. 特使；特刊；专车；特价商品 adj. 特别的 (2B)
specific [spə'sɪfɪk] adj. 特殊的，特定的；明确的；详细的 n. 特性；细节 (2A)
stack [stæk] n. 堆；堆叠 v. 堆积，堆叠 (3A)
standard ['stændəd] n. 标准；水准 adj. 标准的 (1B)
standard ['stændəd] n. 标准，水平 (5L)
starkly ['sta：kli] adv. 严酷地；明显地 (7B)
state [steɪt] n. 国家；州；情形 vt. 规定；声明；陈述 (2A)
stimuli ['stɪmulɪ] n. 促进因素；激励因素 (3L)
storage ['stɔ:rɪdʒ] n. 存储；仓库；贮藏所 (1A)
store [stɔ:] n. 商店；储备 vt. 贮藏，储存 (1A)
structured ['strʌktʃəd] adj. 有结构的；有组织的 (5B)
style [staɪl] n. 风格；时尚 (1A)
suitable ['su:təb(ə)l] adj. 适当的；相配的 (8B)
superiority [sjuˌpiəri'ɔriti] n. 优越，优势 (5L)
supervisory ['sju:pə, vaɪzərɪ] adj. 监督的 (2B)
supplier [sə'plaɪə] n. 供应厂商 (5A)
supply [sə'plaɪ] n. 供给，补给 vt. 供给 (1A)
system ['sɪstəm] n. 制度，体制；系统；方法 (1A)

T

tablet ['tæblɪt] n. 平板电脑 (1A)
tackle ['tæk(ə)l] n. 装备；用具；扭倒 vt. 处理 (8B)
tail [teɪl] n. 尾巴；踪迹 (5B)
tamper ['tæmpə] vi. 篡改 (6B)
tangible ['tæn(d)ʒɪb(ə)l] adj. 有形的；可触摸的 n. 有形资产 (1A)
task [tɒ:sk] vt. 分派任务 n. 工作，作业；任务 (2B)
technological [ˌteknə'lɒdʒɪkl] adj. 技术上的；工艺(学)的 (6A)
telecommunication [ˌtelɪkəmju:nɪ'keɪʃ(ə)n] n. 远程通信；无线电通讯 (3A)
template ['templeɪt；-plɪt] n. 模板，样板 (4B)
temporarily ['temp(ə)r(ər)ɪlɪ] adv. 临时地，临时 (1B)

terabyte ['terəbait] n. 太字节；兆兆位(量度信息单位)(7A)
text [tekst] n. [计] 文本 (4A)
three-dimensional ['θridaɪ'menʃənəl] adj. 三维的；立体的；真实的 (7B)
time-line ['taɪmˌlaɪn] n. 时间线；等时线 (4B)
token ['təʊk(ə)n] n. 表征；代币；记号 (3B)
tolerable ['tɒl(ə)rəb(ə)l] adj. 可以的；可容忍的 (7B)
topology [tə'pɒlədʒɪ] n. 拓扑学 (3B)
touchpad ['tʌtʃpæd] n. 触摸屏设备，触摸板 (1B)
track [træk] n. 轨道；足迹 v. 追踪 (5B)
tradeoff ['tred，ɔf] n. 权衡；折中 (6A)
transaction [træn'zækʃ(ə)n] n. 交易；事务 (5A)
transfer [træns'fɜ] n. 转移；传递；过户 vi. 转让；转学 vt. 使转移 (1B)
translate [træns'leɪt] vt. 翻译；转化；解释；转变为；调动 vi. 翻译 (2A)
transmission [trænz'mɪʃ(ə)n] n. 传递；传送；播送 (3A)
trigger ['trɪgə] vt. 引发，引起；触发 n. 扳机；[电子] 触发器 (7A)
trust [trʌst] n. 信任，信赖；责任 v. 信任，信赖 (5B)
tutorial [tju:'tɔ:riəl] n. 导师辅导(时间)，软件教程 (4B)
typically ['tɪpɪkəlɪ] adv. 代表性地；作为特色地 (1B)

U

ubiquitous [ju:'bɪkwɪtəs] adj. 普遍存在的；无所不在的 (8A)
ubiquity [ju:'bɪkwətɪ] n. 普遍存在 (3L)
unauthorized [ʌn'ɔ: θəraɪzd] adj. 未经授权的；未经批准的 (6B)
undesirable [ʌndɪ'zaɪərəb(ə)l] adj. 不良的；不受欢迎的 n. 不良分子 (2B)
unexpectedly ['ʌnik'spektidli] adv. 出乎意料地，意外地 (8A)
unique [ju:'ni:k] adj. 独特的，稀罕的 n. 独一无二的人或物 (5B)
unpredictable [ʌnprɪ'dɪktəbl] adj. 不可预知的；不定的；出乎意料的 (8A)
unstructured [ʌn'strʌktʃəd] adj. 无社会组织的；松散的；非正式组成的 (7B)
update [ʌp'deɪt] vt. 更新；校正，修正；使现代化 n. 更新；现代化 (7A)
up-front ['ʌpfrʌnt] adj. 预先的；坦率的 (8A)
usable ['ju:zəb(ə)l] adj. 可用的；合用的 (1A)
USB abbr. 端口，通用串行总线(Universal Serial Bus) (1B)
usefulness ['ju:sf(ʊ)lnəs] n. 有用；有效性；有益 (1B)
utilities [ju:'tɪlɪtɪz] n. 公用事业；(计) 实用程序 (2B)
utility [ju:'tɪlɪtɪ] n. 实用；公共设施；功用 adj. 实用的；通用的 (8A)

V

value ['vælju:] n. 值；价值；价格；重要性；确切涵义 (2A)
variety [və'raɪətɪ] n. 多样；种类；杂耍；变化，多样化 (1B)

varnish ['vɒ:nɪʃ] n. 亮光漆，清漆（5A）
velocity [və'lɒsəti] n.（物）速度（7B）
vendor ['vendə:] n. 卖主；小贩；供应商（8A）
veracity [və'ræsəti] n. 诚实；精确性；老实（7B）
versatility [ˌvɜ:sə'tɪləti:] n. 多功能性；多才多艺；用途广泛（1A）
version ['vɜ:ʃ(ə)n] n. 版本；译文；倒转术(4B)
via ['vaɪə] prep. 渠道，通过；经由（1A）
virtualization ['vɜ:tʃʊəˌlaɪzeiʃən] n. [计] 虚拟化（8A）
virtually ['və:tʃʊəli] adv. 事实上，几乎；实质上（1B）
visible ['vɪzəbl] adj. 明显的；看得见的；现有的；可得到的 n. 可见物（2A）
visualization [ˌvɪzjʊəlaɪ'zeɪʃən] n. 形象化；可视性（7A）
voice-over ['vɔis，əuvə] n. 画外音；(电影或电视)旁白（4B）
volleyball ['vɒlɪbɔ:l] n. 排球（6L）

W

webcam ['webkæm] n. 网络摄像头（1B）
website ['websaɪt] n. 网站（5A）
weekday ['wi:kdeɪ] n. 工作日（1L）
well-designed ['weldi'zaind] adj. 精心设计的；设计巧妙的（8B）
widespread ['wʌɪdsprɛd] adj. 普遍的，广泛的；分布广的（8A）
wireless ['wʌɪəlɪs] adj. 无线的；无线电的（3A）
workstation ['wərksteɪʃn] n. 工作站（3B）

Phrases and Expressions

A

a combination of　事物的结合；综合（4A）
a credit or debit card　信用卡或借记卡(5A)
a matter of　大约，……的问题（6B）
a series of　一系列的（4A）
a variety of　种种（5L）
access control　限制访问（6B）
access to　接近；有权使用(3A)
add in　添加；把……包括在内（4B）
Adobe Flash　FLASH 动画制作软件（4B）
air crash　空难(3L)

alternating current(AC) electric power	交流电 (1A)
alternative to	可供选择 可供选择的替代 (4B)
an enormous number of	大量的(3A)
Application Layer	应用层(3A)
application software	应用软件 (2A)
arrived at	到达 (1L)
as follows	如下 (6A)
as is known to all	众所周知(5A)
as shown in	如……所示 (3B)
as well as	也；和……一样；不但……而且(3A)
assembly language	(计) 汇编语言 (2A)
at least 至少	(1B)
at the meantime	同时(5A)

B

B2B(Business to Business)	企业间电子商务(5A)
B2C(Business to Customer)	企业对消费者电子商务(5A)
B2G(Business to Government)	企业与政府机构间电子商务(5A)
back up	支援，备份 (6B)
background music	背景音乐 (4B)
backup copies	副本，备份件 (6B)
based on	基于 (2B)
be capable of	能够(3A)
be concerned with	涉及，参与 (6B)
be connected to	与……有联系；与……连接 (3B)
be considered as	被认为…… (5B)
be regarded as	被认为是，被当作是 (4A)
be replaced with	替换为，以……代替 (4A)
before delivery	交货前(5A)
big data	大数据 (7A)
binary value	二进制数值 (2A)
boot loader	(计) 引导程序 (2B)
Braille embosser	盲人点字浮雕器 (1B)
break into	闯入；破门而入(5A)
build in	插入；嵌入 (4B)
bus network topology	总线网络拓扑结构 (3B)
Business Intelligence	商业智能 (7B)
but not limited to	但不限于 (4A)

C

C2C(Consumer To Consumer)	消费者间电子商务(5A)
capital expenditure	基本建设费用 (8B)
capturing data	数据采集 (7A)
carry out	执行，实行；贯彻；实现 (2A)
cash register	收银处 (2L)
central computer	中央计算机 (3B)
clamshell form factor	翻盖形式 (1A)
clipper chip	加密芯片 (6B)
cloud computing	云计算 (8A)
combat crime	治理罪犯 (7A)
come down to	归根结底，可归结为 (4B)
come over	过来 (2L)
communications protocol	通信协议(3A)
computer case	电脑机箱 (1A)
computer data storage	计算机数据存储器 (1B)
computer network	计算机网络(3A)
computer-based training courses	计算机辅助培训 (4A)
connecting lines	连接线；外线 (3B)
consists of	由……构成；包括 (2A)
conversion value	转换价值(5B)
credit card	信用卡(5L)

D

data link	数据(自动)传输器(3A)
data network	数据网络(3A)
data set	数据集 (7A)
DC power	直流电 (1A)
deal with	处理 (7A)
depend on	依赖于 (1B)
device driver	(计) 设备驱动程序 (2B)
devise ways to	想办法 (6B)
differ in	不同在；在……方面存在不同 (3A)
differ from	与……不同；区别于…… (4B)
digital audio	数字音频(3A)
digital electronics	数字电子技术(3A)
digital footprint	数据痕迹 (7B)
digital video	数字视频(3A)

distinguish... from	与……区别开(5B)
divide into	把……分成 (2B)
do damage to	损坏，损害 (6B)
due to	由于 (1A)

E

each other	互相，彼此 (2A)
either... or	或者，或者 (1A)
electricity network	电网 (8A)
enable... to	使……能够做……(5B)
end user	最终用户(5A)
Ethernet Protocol	以太网协议(3A)
everyday use	日常使用(3A)
expansion bus	系统扩展总线 (1B)
expansion card	扩充插件板 (1B)
expansion slot	扩充插槽 (1A)

F

factor in	将……纳入；把……计算在内 (4B)
fade in	淡入；渐显 (4B)
fight for	为……而战，而奋斗(5B)
fixed media	虚拟硬盘 (1B)
focus on	集中于 (8A)
follow up	跟踪；坚持完成；继续做某事 (4B)
free trial	[贸易] 免费试用 (4B)
French fries	炸薯条 (2L)
from time to time	不时，有时 (6B)
full mesh	全网状 (3B)

G

G2G(Government to Government)	政府间的电子商务(5A)
geometric layout	几何学的布置布线图 (3B)
give attention to	考虑，注意，关心(5B)
go through	参加、经受、通过 (4A)
graphic designer	平面设计师 (4A)

H

hard disk drive (亦作 hard drive)	硬盘驱动器 (1B)
harder and harder	越来越难(5B)

have a good time　玩得开心 (1L)
have access to　使用；接近；可以利用 (4B)
HTTP　超文本传输协议(3A)

I

image scanner　(计) 图像扫描仪 (1B)
in action　在活动；在运转 (4B)
in contrast to　与……对比(或对照) (2A)
in effect　实际上；生效 (3B)
in many instances　在许多情况下 (4B)
in part　在某种程度上 (8B)
in terms of　依据；按照；在……方面 (1B)
in that　因为；由于；既然 (4B)
in the event of　如果，如果……发生，万一 (6B)
in the first place　首先，第一，原本 (6B)
inductive statistic　归纳统计 (7B)
information asset　信息资产 (7B)
information processing　信息加工 (7B)
instant messaging　即时通讯(3A)
instead of　而不是 (8A)
interactive content　互动式内容 (4A)
intermediate node　中间节点 (3B)
international roaming　国际漫游 (3L)
Internet Protocol　互联网协议(3A)
Internet Protocol Suite　互联网协议组(3A)
IP addressing　网络地址(3A)

K

key escrow chip　密钥托管芯片 (6B)

L

lay out　展示；安排；花钱；提议 (3B)
leaf through　翻阅 (7L)
line drawing　素描；线条画 (4B)
link layer　链接层(3A)
linked with　与……有关，与……相连接 (4A)
liven up　使……有生气 (4A)
living standard　生活水平(5L)
Local Area Network　局域网(3A)

logical (or signal) topology	逻辑(或信号)拓扑 (3B)
long-tail keywords	长尾关键词(5B)
look into	调查，观察 (8B)

M

main bus	主总线 (3B)
main unit	主机 (6A)
make a decision	做决定(5B)
make preparations	做准备工作 (6B)
make up	组成；补足；化妆；编造(3A)
making a purchase	购物(5B)
malicious software	恶意软件 (2B)
mesh network topology	网状网络拓扑结构 (3B)
movie clip	影片剪辑 (4B)
multimedia artist	多媒体艺术家 (4A)

N

Network Protocols	网络协议(3A)
network traffic	网络流量(3A)
North Pole	北极 (8L)

O

onone's own	依靠自己 (2A)
on the internet	在网上(5A)
on top	在上面；领先；成功 (2B)
operating system	操作系统 (2B)
optical disc	(计) 光盘 (1B)

P

P2P (peer to peer) voice communication	点对点(对等点)语音通信(3A)
partial mesh	部分网状 (3B)
participate in	参加；分享(5A)
pass through	穿过……；通过…… (3B)
pay as you go	账单到期即付 (8A)
physical security	实体安全 (6A)
physical part	物理部件 (1A)
physical topology	物理拓扑 (3B)
place orders	下订单(5A)
potential customers	潜在客户(5B)

power supply unit 电源(箱)(1A)
practical joke 恶作剧(2B)
prevent... from being 防止……(6A)
printed circuit board 印刷电路板(1B)
processor chip 处理器芯片(6B)
product reviews 产品评论(5B)

R

rather than 而不是(5A)
raw materials 原材料(5A)
refer to 参考;涉及;指的是;适用于(3A)
regardless of 不管,不顾(8B)
removable media 可移动媒介(1B)
rest on laurels 吃老本(5B)
rich snippet 富摘要(5B)
ring network topology 环形网络拓扑(3B)

S

sales path 销售路径(5B)
sales revenue 产品销售收入(5B)
scale up 按比例(8B)
science fiction 科幻小说(4A)
search traffic 搜索流量(5B)
security buffer 安全缓冲区(6B)
sense of touch 触觉(4A)
SEO(Search Engine Optimization 搜索引擎优化(5B)
service providers 服务供应商(5B)
service-oriented 服务导向,服务至上(8A)
set you back 让你破费(4B)
shout out 大叫,大声喊(3L)
slide shows 放映幻灯片(4A)
SMTP 简单邮件传输协议(3A)
so forth 等等(6A)
special effects 特技效果(4A)
specialize in 专营,专门研究,专攻(5B)
spot business trend 发现市场趋势(7A)
stand to lose 一定失利(6A)
star network topology 星形网络(3B)
storage devices 存储设备(6A)

storage server	存储服务器(3A)
structured data	结构化数据(5B)
such as	像 (1A)
supply chain	供应链(5A)
system software	系统软件 (2B)

T

take place	发生(5A)
take this example	举个例子来说(5B)
take turns	依次；轮流 (2L)
tampered with	篡改 (6B)
this is not always the case	情况并非一直如此 (3B)
trade offs	权衡，交易 (6A)
traffic flowing	交通流量(3A)
transmission Control Protocol	传输控制协议(3A)
transmission medium	传送介质(3A)
transport Layer	传输层(3A)
tree network topology	越权存取 (6B)

U

up and running	正常运行 (8A)
URL(Uniform Resource Locator)	全球资源定位器(5B)
USB flash drive	闪存盘，随身碟，优盘 (1B)

V

video compression	视频压缩 (4B)
virtual reality	虚拟现实 (4A)

W

ways of	依靠，方法 (6B)
web browser	网络浏览器 (8B)
weigh against	权衡，与之相当 (6A)
with ease	熟练地，不费力地(5A)
World Wide Web	万维网(3A)

Reference

[1] 董晓霞，孙岩，高炯，等. 计算机英语[M]. 北京：清华大学出版社，2013.

[2] 覃小立，智尚轩，等. 计算机专业英语[M]. 武昌：武汉大学出版社，2010.

[3] 蒋秉章，等. 新航标职业英语[M]. 北京：北京语言大学出版社，2015.

[4] 胡英坤，车丽娟，等. 商务英语写作[M]. 北京：外语教学与研究出版社，2005.

[5] 赵萱，郑仰成，等. 实用英语应用文写作教程[M]. 北京：高等教育出版社，2013.

[6] 苏兵、张淑荣. 计算机英语[M]. 2 版. 北京：化学工业出版社，2014

[7] Tay Vaughan. Multimedia Making it Work. Computer. OH：The McGraw-Hill Company，2000.

[8] Douglas E Comer. Computer Networks and Internet. New York：Prenfice Hall Ine.，1998.

[9] https：//www. wikipedia. org/

[10] https：//www. linkedin. com/feed/? trk＝brandpage_360_def-mainlink

[11] https：//www. presentationmagazine. com/how-to-create-a-multimedia-presentation-164. htm